地质学专业英语
ENGLISH FOR GEOLOGY

朱海丽　胡夏嵩　主编

图书在版编目(CIP)数据

地质学专业英语/朱海丽,胡夏嵩主编. —武汉:中国地质大学出版社,2024.9—ISBN 978-7-5625-5984-9

Ⅰ. P5

中国国家版本馆 CIP 数据核字第 2024XS7637 号

地质学专业英语		朱海丽　胡夏嵩　主编
责任编辑:唐然坤	选题策划:唐然坤	责任校对:胡　萌
出版发行:中国地质大学出版社(武汉市洪山区鲁磨路388号)		邮编:430074
电　　话:(027)67883511	传　　真:(027)67883580	E-mail:cbb@cug.edu.cn
经　　销:全国新华书店		http://cugp.cug.edu.cn
开本:787毫米×1092毫米　1/16		字数:184千字　　印张:6
版次:2024年9月第1版		印次:2024年9月第1次印刷
印刷:武汉市籍缘印刷厂		
ISBN 978-7-5625-5984-9		定价:25.00元

如有印装质量问题请与印刷厂联系调换

前　言

专业英语教学是大学英语教学中不可或缺的组成部分，对培养学生的文献阅读能力、科技论文写作能力非常重要。本教材为编者在多年来为地质类专业应用型本科生讲授专业英语课程的教学实践经验基础上，并汲取各兄弟院校教学改革成果和国内各类专业英语教材的优点编写而成，目的是为地质类专业应用型本科生提供一本难度适中的实用教材。

《地质学专业英语》适用于地质类资源勘查与地质工程专业四年制本科学生教学。全书包括 7 个部分共 12 课内容：第一部分为地质学基础知识内容(Lesson One)；第二部分为矿物学方面内容(Lesson Two)；第三部分为岩石学方面内容(Lesson Three to Lesson Five)；第四部分为古生物学与地层学方面内容(Lesson Six)；第五部分为构造地质学方面内容(Lesson Seven to Lesson Nine)；第六部分为水文地质学与工程地质学方面内容(Lesson Ten and Lesson Eleven)；第七部分为英文摘要的写作技巧与方法(Lesson Twelve)。本教材设有课文和相应阅读辅助材料共 22 篇，阅读材料主要选自地球科学相关原文书刊并根据最新英语使用规范略微修改；每篇课文后面均附有词汇供学生参考学习；此外，每篇课文后附有科技英语翻译方法与技巧，可供学生学习。考虑到应用型本科教学的特点，本教材内容通俗易懂、可读性强；通过介绍地质领域相关的基础知识和成果，让学生在英语环境中学习地球科学知识，掌握基本专业术语，提高专业文献阅读能力；了解本专业英文科技写作特点和科技论文英文摘要写作特点、写作方法与基本写作技巧等方面的基础知识，从而提升学生的专业英语阅读能力和翻译水平。

本教材根据教育部《普通高等学校本科专业目录(2024 年)》的基本要求和青海大学最新修订的本科生培养方案编写而成。编者参阅了国内外出版的一些同类教材及网络资料，于 2017 年 7 月在青海大学教务处教材资金的资助下完成，作为校内教材已使用多年，并经多次修改和补充。

在编写的过程中，编者得到了青海大学教务处的支持和帮助，在此对他们及对本教材所引用文献的著作者表示衷心的感谢。

因编者水平有限，书中不足之处在所难免，敬请各位读者批评指正。

朱海丽于西宁
2024 年 7 月

Contents

Lesson One Earth Sciences ……………………………………………………… (1)
    Phrases and Expressions ……………………………………………………… (2)
    Exercises ……………………………………………………………………… (3)
    Translation Tips ……………………………………………………………… (4)
    Reading Material ……………………………………………………………… (5)

Lesson Two Minerals ……………………………………………………………… (7)
    Phrases and Expressions ……………………………………………………… (8)
    Exercises ……………………………………………………………………… (8)
    Translation Tips ……………………………………………………………… (10)
    Reading Material ……………………………………………………………… (11)

Lesson Three Igneous Rocks …………………………………………………… (13)
    Phrases and Expressions ……………………………………………………… (14)
    Exercises ……………………………………………………………………… (14)
    Translation Tips ……………………………………………………………… (16)
    Reading Material ……………………………………………………………… (18)

Lesson Four Fragmental Rocks ………………………………………………… (19)
    Phrases and Expressions ……………………………………………………… (20)
    Exercises ……………………………………………………………………… (20)
    Translation Tips ……………………………………………………………… (21)
    Reading Material ……………………………………………………………… (23)

Lesson Five Layering or Metamorphic Rocks ………………………………… (24)
    Phrases and Expressions ……………………………………………………… (25)
    Exercises ……………………………………………………………………… (25)
    Translation Tips ……………………………………………………………… (27)
    Reading Material ……………………………………………………………… (29)

Lesson Six Fossils and How They Are Made ………………………………… (30)
    Phrases and Expressions ……………………………………………………… (31)
    Exercises ……………………………………………………………………… (31)

    Translation Tips .... (33)
    Reading Material .... (34)
Lesson Seven Description and Classification of Joints .... (36)
    Phrases and Expressions .... (37)
    Exercises .... (38)
    Translation Tips .... (39)
    Reading Material .... (41)
Lesson Eight Recognition of Faults .... (43)
    Phrases and Expressions .... (44)
    Exercises .... (45)
    Translation Tips .... (46)
    Reading Material .... (49)
Lesson Nine Parallel Folds .... (51)
    Phrases and Expressions .... (52)
    Exercises .... (53)
    Translation Tips .... (54)
    Reading Material .... (55)
Lesson Ten Ground Water .... (56)
    Phrases and Expressions .... (57)
    Exercises .... (58)
    Translation Tips .... (59)
    Reading Material .... (61)
Lesson Eleven Soil Composition .... (63)
    Phrases and Expressions .... (64)
    Exercises .... (64)
    Reading Material .... (66)
Lesson Twelve Writing .... (67)
    The Characteristics of Scientific and Technological English Writing .... (67)
    The Skills and Methods for English Abstract Writing .... (68)
    Exercise .... (73)
Appendix I Words and Expressions .... (75)
Appendix II Plate .... (81)

Lesson One
Earth Sciences[①]

【1】Such then, in brief, is the ever changing planet upon which we dwell. The study of this planet is particularly the province of what we call the earth sciences. The basic earth science is geology (from the Modern Latin geo- "earth" and -logia "science"). It examines the different processes by which the rock structures and the landscape of the earth have come into being and the present composition of the earth—it's exterior and interior. Geology also reconstructs the sequence of past changes in the earth's structure. This leads to an understanding not only of the history of the earth itself, but also of the different forms of life that dwelt upon it in past ages.

【2】Geology has a number of subdivisions. The science of **petrology** studies the character and origin of all kinds of rocks. Structural geology examines the structural arrangements of rocks, brought about as a result of folding and fracturing. The minerals that make up rocks are studied in the science of **mineralogy**. **Stratigraphy** deals with the sequence of rock layers, particularly sedimentary rock. **Geomorphology** considers the various land forms that have been sculptured by such surface agencies as wind, flowing water, and precipitation. **Glaciology** deals with the processes connected with ice and snow. **Paleontology** studies the plant and animal fossils buried in the earth and the evidences of the gradual development of life in

petrology n. 岩石学
mineralogy n. 矿物学
stratigraphy n. 地层学
geomorphology n. 地貌学
glaciology n. 冰川学
paleontology n. 古生物学

---

①苏生瑞. 地质专业英语[R]. 唐山：唐山工程技术学院，1986(内部资料).

past ages. The rock deposits that are valuable to man (metals, coal, petroleum, and the like) are the subject matter of economic geology. Exploration for valuable deposits and their removal is called mining geology. Civil engineers must know about the qualities and strength of the earth and rocks they encounter and also use in their constructions, this study is engineering geology.

【3】 Certain other sciences may be included under geology by some experts. **Geophysics** deals with the physics of our planet. **Geochemistry** treats of the composition of the earth's rocks, minerals, soils, air, and water. The soil itself belongs under the science of **pedology**. **Hydrology** takes in all the bodies of water-lakes, streams, and swamps on the landmasses of the world. Marine, or submarine geology (geological oceanography) is concerned with the sea bottom and **shorelines**. The exact **delineation** of the earth's features is the concern of **topography**, **hydrography** (for water bodies), and **mapmaking**.

【4】 Fields closely related to geology are **physiography** and physical geography, dealing with landscapes in different countries; **ecology** and **biogeography**, concerned with life in relation to its environment and to the land and sea; **limnology**, with the suitability of water for life in it; **oceanography**, the science of the sea; **meteorology**, the study of weather; climatology, the study of climates; and **astrogeology** (planetary astronomy), dealing with the nature of planets, moons, **asteroids**, and **meteorites**.

geophysics   n. 地球物理学
geochemistry n. 地球化学
pedology   n. 土壤学
hydrology   n. 水文学
shoreline   n. 海岸线
delineation   n. 描写，描绘
topography   n. 地形学
hydrography   n. 水文地理学
mapmaking   n. 制图学
physiography n. 地文学，自然地理学
ecology   n. 生态学
biogeography n. 生物地理学
limnology   n. 湖沼学
oceanography n. 海洋学
meteorology   n. 气象学
astrogeology n. 天体地质学
asteroid   n. 小行星
meteorite   n. 陨星，陨石

Phrases and Expressions

structural geology   构造地质学
economic geology   经济地质学
mining geology   矿山地质学
engineering geology   工程地质学

marine (submarine) geology　海洋(海底)地质学
planetary astronomy　行星天文学

Exercises

I. Translate the following sentences into English.

1. 简言之,地学就是研究地球的科学。

2. 古生物学要重建生物演化的序列。

3. 地质学是地学的一个分支。

4. 矿物学是岩石学的基础。

II. Answer the following questions.

1. Where does "geology" come from?

2. What does geology study?

3. What does structural geology deal with?

4. What does geomorphology consider?

5. What does engineering geology study?

Ⅲ. Choose an appropriate word from the following box to fill in each of the following blanks. Each word can be used only ONCE. Change the form where necessary.

| although | experiment | seek | change | over |
| chemistry | explain | indoor | conflict | volcano |

Geology is based mainly on observations and **seeks** to determine the history of the earth by **explaining** these observations logically, using other sciences such as physics, **chemistry**, and biology. Only a small part of geology can be approached **experimentally**. For example, **although** the important use of fossils to date or establish contemporaneity of rock strata is based on the simple, basics principle that life has **changed** during the history of the earth, this principle could not be established experimentally; it was the result of careful observations and analyses **over** a long period of time by many people of varied backgrounds.

Geologic problems are many, diverse, and complex; almost all must be approached indirectly. And in some cases, different approaches to the same problem lead to **conflicting** theories. It is generally difficult to test a theory rigorously for several reasons. The scale of most problems prevents laboratory study; that is, one cannot bring a **volcano** into the laboratory, although some facets of volcanoes can be studied **indoors**. It is also difficult to simulate geologic time in an experiment. All of this means that geology lacks exactness and that our ideas change as new data become available. This is not a basic weakness of geology as a science, but means only that much more remains to be discovered; this is a measure of the challenge of geology.

Translation Tips

Ⅰ. 翻译标准

翻译必须达到"信""顺"。"信"即忠实于原作,"顺"即译文通顺,合乎现代汉语的规范。

忠实于原作指忠实于原作的内容,对原作所述内容不增添和删减,不歪曲和改动,同时还要保持原文风格。

译文要合乎现代汉语的规范,就要用现代的、通用的语言,遣词造句要符合汉语语法和汉语表达习惯,不生搬硬套不符合汉语习惯的原文句式和搭配。

Ⅱ. 翻译过程

翻译过程大体分为理解与表达两个阶段。理解是指通过原作的语言形式理解原作的内容,表达是指在译作语言中寻找和挑选恰当的表达方式。

理解是表达的前提和基础,不理解谈不上表达,理解不准确就会产生错误的表达。

例:Both of the substances do not dissolve in water.

这两种物质都不溶于水。(误)

这两种物质并不都溶于水。(正)

上面的错误译句便是没有准确理解原文的结果。显然,翻译科技作品本来是为了传播知识,而如果把关键的地方,如定义、结论等翻译错了,就起不到传播知识的作用。

表达就是把理解具体化。英、汉两种语言相互之间存在着差异,同一个英文句子译为汉语时可以有若干种表达方式,这就要根据汉语表达习惯的要求和上下文的内在联系,寻找和挑选最恰当的表达方式。

例:Some atoms are so constructed that they lose electrons easily.

某些原子是如此被构造的,以致它们很容易失去电子。

上述译文大体上忠实于原作的内容,但语言生硬,不合汉语表达习惯,不如下面的译文表达恰当:某些原子的结构使它们很容易失去电子。

理解和表达是一个统一的过程,正确的理解和恰当的表达是达到翻译标准的前提。

Reading Material

Something about Geology[②]

Geology, the science of the earth, tells us about the world, what it is made of, its age, and its aging processes, its land forms, and its abyss ocean depths. It traces life from the first sea-spawned vegetation and animals, through dinosaurs, to man—all on a majestic time scale that staggers the imagination.

These secrets of the earth are probed in many ways. Waves from earthquakes travel through the interior to bring out messages about its structure that are written on seismographs. Chemical and X-ray analyses reveal the composition of rocks and minerals. Mapping of sands, gravels, and boulders shows where glaciers once moved

---

[②] FOSTER R J. General geology[M]. Columbus, OH: Charles E. Merrill Publishing Company, 1983.

over the land. Examination of ancient caves and strand lines proves that the oceans have been hundreds of feet deeper than they are today, while submarine canyons and wave-planed volcanic peaks under deep water present the possibility that the oceans have also been hundreds, even thousands, of feet shallower.

Geology is all this, and much more. It is a synthesis of the natural sciences: astronomy, biology, chemistry, mathematics, and physics. And, above all, it has something for everyone. Who can experience or even hear about an earthquake or volcanic eruption without wondering about its cause? If you found a sea shell of fish solidly encased in the rock of an inland stream bed, or of a high mountain, would you wonder why it was there? Have you ever pondered the jumbled varicolored rocks or multitudinous grains of sand of a shoreline, the gold like glitter of yellow mica in a piece of field stone, or the smooth symmetry of a quartz crystal? If these or any of thousand and one phenomena all around us have stimulated so much as a fleeting question in your mind, you have peeked through a door into the world of geology. Anyone can walk through such a door and find treasures limited only by the dimensions of his curiosity and enthusiasm.

Lesson Two
Minerals[①]

【1】 Elements combine to form **minerals**. In view of the age-old dependence of man on minerals for his weapons, his **comforts**, his **adornments**, and often for his **pressing** needs, it is surprising that many persons have only a **vague** idea about the nature of a mineral. Yet anyone who has climbed a mountain walked on a sea **beach**, or worked in a garden has seen minerals in their natural **occurrence**. The rocks of the mountain, the sand on the beach, the soil in the garden are completely or in large part made up of minerals. Even more familiar in everyday experience are products made from minerals, for all articles of commerce that are inorganic, if not minerals themselves, are mineral in origin. All the common building materials such as steel, cement, brick, glass, and **plaster** had their origin in minerals.

【2】 Minerals are products of natural processes. The greatest bulk of minerals occur as essential and **integral** constituents of rocks, others are frequently found in **veins** and cavities. **Mineralogists** limit the term "mineral" to include only those materials which occur naturally. Thus, steel, cement, plaster, and glass, although all are derived from naturally occurring mineral raw materials are not regarded as minerals themselves, since they have been processed by man. Excluded, also are all substances resulting directly from the

mineral   n. 矿物
comfort   n. 舒适,使舒适的事物
adornment   n. 装饰,装饰品
pressing   adj. 紧迫的,迫切的
vague   adj. 模糊的
beach   n. 海(河、湖)滩
occurrence   n. 出现,产状
plaster   n. 灰泥,熟石膏
integral   adj. 构成整体所必需的,完整的
vein   n. 脉,岩脉
mineralogist   n. 矿物学家

---

① 苏生瑞. 地质专业英语[R]. 唐山:唐山工程技术学院,1986(内部资料).

processes of plant and animal life. Thus coal, oil amber, and bones of animals are excluded, even though they occur naturally in the earth's crust. The most important and significant limitation placed on the definition of a mineral is that it must be an element or a compound of elements that can be expressed by a chemical formula. This means that a specific mineral is always composed of the same elements in the same proportions.

pyrite *n.* 黄铁矿
sulfur *n.* 硫
frame   *v.* 构成，限定

【3】 If a mineral is composed of one element, its formula is merely the symbol of the element, as Au for gold. If the compositions of other minerals are to be expressed by formulas, the elements that make them up must always combine with one another in fixed simple ratios. For the common mineral quartz, the ratio is one atom of silicon to two of oxygen, with the formula SiO_2 and for the mineral **pyrite**, the formula FeS_2 indicates the ratio of one atom of iron to two of **sulfur**.

【4】 Now that we have determined what will be included and what excluded, we can **frame** a definition of a mineral as a naturally occurring element formed as a product of inorganic processes.

Phrases and Expressions

in view of   由于……，鉴于……，考虑到
age-old   古老的，久远的

Exercises

Ⅰ. **Translate the following sentences into English.**

1. 某些矿物与水化合后形成其他较复杂的矿物。

Lesson Two　Minerals

2. 地质学不局限于探矿和研究矿物，而包括广阔得多的领域。

3. 在造岩矿物中最丰富的元素是硅、氧、铝、铁、钙等，所有其他元素，甚至大家熟知的铜、铅、锌和氮都是少量的。

4. 矿物在地球内部的高温高压条件下可能变得更加致密。

Ⅱ. **Answer the following questions.**

1. Why do we say that minerals are familiar to every one?

2. What is meant by the term "mineral"?

3. Are coal, oil and amber minerals? If not, why?

4. What is the most important limitation placed on the definition of a mineral?

5. How do we express minerals chemically? Give some examples of common minerals.

Ⅲ. **Choose an appropriate word from the following box to fill in each of the following blanks. Each word can be used only ONCE. Change the form where necessary.**

| crystal | bound | break down | inorganic | oppose |
| order | material | geometry | small | combine |

So far we have discussed elements, which are _____ that cannot be subdivided by ordinary chemical methods. The _____ unit of an element is the atom. The atom can be _____ into smaller particles such as electrons, neutrons, protons, etc., but except in the case of radioactive elements, this requires the large amounts of energy available in atom-smashing machines such as the cyclotron.

In nature, elements are _____ to form minerals, which can be defined as:

1. naturally occurring, crystalline;

2. _____ substances with;

3. a definite small range in chemical composition and physical properties.

Note that all three conditions must be met by a mineral.

Minerals are _____ substances. That is, they have an _____ internal structure (arrangement of atoms), as _____ to such things as glasses which are super-cooled liquids and have no internal order. A crystal is a solid form _____ by smooth planes which give an outward manifestation of the orderly internal structure. Note the distinction between crystal and crystalline. Although all minerals are crystalline, they do not necessarily occur as _____ crystals.

Translation Tips

Ⅰ. 词义的确定

1. light

Since natural gas is **lighter** than oil, the gas accumulates above the oil.（形容词）

Sound travels much more slowly than **light**.（名词）

Our houses are **lighted** by electricity.（动词）

2. well

The **well** is deep enough to reach the ground water table.（名词）

In the early 1960s, it was suggested that the ridges might mark areas where material was **welling** up from inside the earth.（动词）

The cone by then was **well** over 450 feet high.（副词）

Ⅱ. 词义的表达

1. build(动词)

build a house	build a ship	build a bridge
build a dam	build a fire	build a railway
build socialism		

2. heavy(形容词)

heavy mineral	heavy roads	heavy objects
heavy oil	heavy rain	heavy machine
heavy crops	heavy works	heavy storm

Lesson Two Minerals

heavy current heavy fire heavy traffic
a heavy sea a heavy sky

The facts and theories **at its core** are the **deposit** of a good three centuries of hard work and hard thinking.

The roar and crash of **man-made structures mingled** with a dull booming from the earth itself.

A stream's velocity is its rate of flow, **measured** in feet per second or miles per hour.

Gathering facts, confirming them, suggesting theories, testing them, and organizing **findings**—this is all the work of science, and the method of carrying it out are sometimes brought together and **labeled** the scientific method.

Reading Material

Common Minerals: Feldspar, Quartz, and Mica[2]

Feldspar

The most abundant mineral type, feldspar, composes over 60% of the rock materials in the earth's crust. Strictly speaking, the term feldspar refers to a group of closely related minerals having generally similar composition and characteristics. They are alumina-silicates of sodium, potassium, and calcium, which explain why these elements, along with oxygen, silicon, and alumni, are so abundant in the earth's crust. Potassium feldspar includes the minerals orthoclase and microcline. Plagioclase includes several sodium and calcium bearing feldspars.

---

[2] FOSTER R J. General geology[M]. Columbus, OH: Charles E. Merrill Publishing Company, 1983.

Quartz

A very widespread mineral, quartz is the second-most abundant. Quartz is a specific mineral, the only common one of the silica group. Chemically, silica is SiO_2; thus quartz is a compound composed entirely of the two most abundant chemical elements. In large pure crystals, quartz resembles colorless glass. However, slight impurities may give it a variety of colors, and some minutely crystalline varieties, such as flint, may be opaque and of a waxy luster.

Mica

Mica is the name of a group of minerals which are readily split into thin flexible sheets. This distinctive property results from a single, perfect, cleavage plane, which is repeated throughout the mineral. The minerals are very complex potassium, alumina-silicates with added oxygen-hydrogen combinations. Two common minerals in this group are dark mica (biotite), which also contains iron and magnesium, and colorless or white mica (muscovite), in which these two elements are absent. The atoms of all these elements are arranged in a complicated manner to produce a loosely bonded sheet-like structure within the crystal. This is responsible for the characteristic cleavage.

Lesson Three Igneous Rocks

【1】 The molten, **seething** mass from which the igneous rocks have been formed is called **magma**. It contains various gases in solution, but for the most part it is made up of rock-forming minerals, dissolved in **haphazard** fashion.

【2】 According to some theories of the earth's formation, our planet was once a mass of **swirling** gases cast off from the sun. As these gases gradually cooled, they were converted into liquid form. This liquid—the magma—began to cool and the minerals it contained began to **crystallize**. Heavy minerals tended to sink in the still-liquid mass of magma. Lighter minerals floated on top of the heavier ones. As the cooling continued, the rocks began to **solidify**. In time they formed a solid crust.

【3】 This represented only one stage in the formation of igneous rock. Vast quantities of magma at extremely high temperatures were **imprisoned** under the crust. The gases contained in the magma **exerted** enormous pressures not only upon the igneous rocks that already been formed but also upon an entirely different kind of formation—the sedimentary rocks.

【4】 The original igneous rocks of the earth's crust and the sedimentary rocks that were formed later were sometimes unable to resist the pressure of the magma under the crust. The liquid rock would then make its way up to higher levels.

seething *adj.* 炽热的,沸腾的
igneous *adj.* 火的,火成的
magma *n.* 岩浆
haphazard *adj.* 偶然性,杂乱的
swirling *adj.* 旋转的
crystallize *v.* 使结晶
solidify *v.* 使固化,使凝固
imprison *v.* 束缚,堵塞
exert *v.* 用(力)

---

① 苏生瑞. 地质专业英语[R]. 唐山:唐山工程技术学院,1986(内部资料).

【5】 In some cases the advance of magma was stopped by the rock layers in its path—only its gases escaped. The magma gradually cooled and solidified under the rock layers that made up the upper part of the earth's crust. Igneous rocks that have been formed in this way are called **intrusive** masses.

【6】 In other cases the magma failed to be **checked** in its upward progress by the surrounding rocks. It reached the surface of the earth and was **discharged** either through a simple **opening**, or volcanic vent, or through a crack in the rocks. Igneous rocks formed from magma **extruded**, or **thrust out**, from the earth are known as **extrusive** masses.

intrusive　*adj.* 侵入的
check　*n.* 制止,控制
discharge　*v.* 排泄,释放
opening　*n.* 孔,空隙
extrude　*vt.* 挤压出,喷出,突出
extrusive　*adj.* 喷出的

Phrases and Expressions

igneous rock　火成岩
rock-forming　造岩的
cast off　抛弃
convert into　转变成
heavy mineral　重矿物
still-liquid　仍处于液体的
in time　及时,按时,准时
intrusive mass　侵入岩
surrounding rock　围岩
volcanic vent　火山口
thrust out　挤出,接出
extrusive mass　喷出岩

Exercises

Ⅰ. Translate the following sentences into English.

1. 岩浆是含有水、气体等物质的熔融状态的物质。

Lesson Three Igneous Rocks

2. 岩浆可以在地下冷却也可以在地表冷却,在地下冷却形成的岩石叫侵入岩。

3. 岩浆的冷却是一个复杂的物理化学过程。

II. Answer the following questions.

1. What is magma?

2. What does magma contain mainly?

3. What is the effect of gases contained in the magma?

4. Describe the cooling process of magma.

III. Choose an appropriate word from the following box to fill in each of the following blanks. Each word can be used only ONCE. Change the form where necessary.

```
porous    eruption    later    composite    volatile
ocean     minor       exist    active       carbon dioxide
```

A great deal can be learned about igneous processes from the study of volcanoes. They are, after all, the only direct evidence we have for the _____ of magma in the crust.

The _____ volcanoes today are of several different types, and the differences among them seem to depend on the _____ of their magmas, more particularly on the behavior of the _____ in the magmas. The most abundant volatile in magma is water, which escapes from volcanoes in the form of steam. _____ is a common volcanic gas, but the sulfur gases (hydrogen sulfide and the oxides of sulfur), because of their strong odors, are the gases most easily noted near volcanoes. In addition, _____ amounts of other gases, such as carbon monoxide, hydrochloric and hydrofluoric acid, ammonia, hydrogen, and hydrocarbons, are also released. Ordinary air also escapes from some volcanoes, especially those with _____ rocks. Volcanic emanations are believed to have played a major role in the formation of the _____. and the atmosphere. Recent _____ at Hawaii are estimated to release 1-2 percent gas during the early stages of eruption and the early stages of eruption and about one-half percent during _____ stages. Laboratory studies suggest that at depth, magmas may contain up to 5-8 percent dissolved water.

Table 1 Composition of volcanic gas at Hawaii

Volcanic gas	Water	Carbon dioxide	Hydrogen	Nitrogen	Argon	Sulfur dioxide	Sulfur trioxide	Sulfur	Chlorine	Total
Volume/%	70.75	14.07	0.33	5.45	0.18	6.40	1.92	0.10	0.05	99.65

Translation Tips

英译汉时，为了适应汉语的表达习惯，要将句中的某些词的词性加以改变，这叫作词性转换。常见的有以下几种。

Ⅰ. 名词转换为动词

1. 由动词派生成转化的名词

A change of state from a liquid to a gas form requires heat energy.

Some scientists suggested that there had been a gradual evolution of life on the earth.

2. 某些同介词组合在一起的名词

In their study of rocks and minerals, geologists apply chemical and physical concepts of atoms, molecules, and crystals.

Clouds are formed by the evaporation of the water in the atmosphere.

Ⅱ. 动词转换为名词

英语中有些动词在汉语中没有相应的动词形式，碰到这类动词汉译时通常都换成名词。

The earth on which we live is shaped like a ball.

A highly developed physical science is characterized by an extensive use of mathematics.

Ⅲ. 形容词转换为其他词

1. 转换为副词

当形容词所修饰的名词译为动词时，形容词自然被译成副词，前面已有这样的例子。

Geology is concerned with systematic study of rocks and minerals.

Since the wide application of computers in the late 1950s, geologists have been increasingly attracted to mathematical methods of data analysis.

Only when we give full play to man's initiative can we make full use of machines to transform nature.

2. 转换为动词

Natural gas is often present in the reservoir rock.

Both of the substances are not soluble in water.

3. 转换为名词

Evidence indicates that the outer core is about twice as dense as the material in the mantle.

Glass is much more soluble than quartz.

4. 副词转换为形容词

Earthquakes are <u>closely</u> related to faulting.

The magma within the earth may be <u>heavily</u> charged with gases and steam.

Ⅳ. 介词转换为动词

介词所连接的词之间有时含有动作意味,因此有时可将介词译成动词。

There is no generally accepted name <u>for</u> the rock layers formed during an era.

Ground water also contains dissolved carbon dioxide from the air and from the exchange of gases <u>in</u> plant roots.

Reading Material

Minerals in Igneous Rocks[②]

The minerals that have contributed to the formation of igneous rocks are rather limited in number and in variety. Commonest of all is quartz, a very hard mineral which is a compound of silicon and oxygen. The feldspars are also abundant. They are light to dark in color and contain potassium, sodium, or calcium, as well as aluminum, silicon, and oxygen. Pyroxene and hornblende, containing different metals plus silicon and oxygen, are darker and heavier. The micas are sheet-like minerals composed of aluminum, other metals, oxygen, and silicon. Magnetite, an iron-oxygen compound, is heavy and magnetic. Olivine, consisting of iron, magnesium, oxygen, and silicon, is a green mineral.

---

②FOSTER R J. General geology[M]. Columbus, OH: Charles E. Merrill Publishing Company, 1983.

Lesson Four
Fragmental Rocks

【1】 The sedimentary rocks that made up entirely of particles of other rocks are known as **fragmental**, or **clastic** rocks (clastic comes from the Greek klastos, meaning "broken"). The **fragments** from which the fragmental rocks are derived are generally classified, on the basis of size, in four groups: **gravel**, sand, **silt**, and mud.

【2】 Gravel is the **coarsest** sediment of all. It consists of fragments that are at least 2 millimeters in diameter. Gravel fragments range from small **pebbles** to big **boulders**. In between there are the particles called **cobbles**. Sand sediments consist of rounded grains ranging from 2 millimeters in diameter to 1/16 millimeter. Particles smaller than 1/16 millimeter and at least 1/256 millimeter in diameter are called silt. The finest sediments—mud and **clays**—consist of fragments under 1/256 millimeter in diameter.

【3】 When more or less rounded gravel particles—pebbles, cobbles, and boulders—are cemented together, they form the type of rock called **conglomerate**. In the variety known as **breccias**, the gravel fragments are not rounded, but angular.

【4】 Cemented sand grains give rise to the **porous** formation known as **sandstone**. The **pores** of this rock may make up as much as 30 percent of the total volume. Liquids move quite freely through the pores. For this reason, sandstones

fragmental   *adj.* 碎屑的，破碎的
clastic   *adj.* 碎屑状的
fragment   *n.* 碎片，碎屑
gravel   *n.* 砾石，砂砾
silt   *n.* 粉砂
coarse   *adj.* 粗粒的
pebble   *n.* 细砾，卵石
boulder   *n.* 巨砾
cobble   *n.* 大砾，卵石，圆石
clay   *n.* 黏土
conglomerate   *n.* 砾岩
breccia   *n.* 角砾岩
porous   *adj.* 松散的
sandstone   *n.* 砂岩
pore   *n.* 孔，孔隙

---

① 苏生瑞. 地质专业英语[R]. 唐山：唐山工程技术学院, 1986(内部资料).

are often reservoirs for petroleum deposits, as well as for **ground water**. Silt particles are converted into **siltstone**; mud particles, into **mudstone** and **shale**.

siltstone　　n. 粉砂岩
mudstone　　n. 泥岩
shale　　n. 页岩

Phrases and Expressions

sedimentary rock　　沉积岩
ground water　　地下水

Exercises

Ⅰ. Translate the following sentences into English.

1. 火成岩占地壳的 95% 左右。

2. 碎屑岩是沉积岩的一种。

3. 在沉积岩中能找到化石。

4. 根据砾石排列方向可以推断河流流向。

Ⅱ. Answer the following questions.

1. What is the most important factor on which the classification of fragmental rocks is based?

2. How does conglomerate form?

3. How much may the pores of sandstone make up?

4. Arrange the fragmental rocks by the size of fragments.

Ⅲ. Choose an appropriate word from the following box to fill in each of the following blanks. Each word can be used only ONCE. Change the form where necessary.

> form   lithify   weather   sedimentary rocks   clay
> silica   iron oxides   calcite   well-sorted   cement

The products of _____ generally are eroded, transported, and deposited before they are transformed into _____. The transformation of a sediment into a rock is called _____. Several processes are involved. _____ is the deposition by ground water of soluble material between the grains and, as might be expected, is most effective in coarse-grained, _____, permeable rocks. The main cementing agents are _____ recognized by acid test. _____ generally produces the toughest rocks. _____ color the rock red or yellow. Other cements, such as dolomite, are possible. In certain poorly sorted rocks, _____, which generally colors the rock gray or green-grey, may be thought of as _____ the cement.

Translation Tips

Ⅰ. 宾语译作主语

1. 动词宾语译作主语

如果原文句子中的动词宾语在意义上与主语联系密切,或者是主语所代表的事物的某一部分,或者是主语的某种属性等,英译汉时,一般把这种动词宾语译作主语。

Calcite has a relative density of $2.7 g/cm^3$.

A neutron has approximately the same weight as a proton.

2. 介词宾语译作主语

当原句子中介词宾语在意义上是主语所代表的事物的某一属性,或者是主语的某个方面、某一方法或某一位置等时,有时可以把这个介词宾语改译成主语,而把介词省去不翻译。

Mountain glaciers may vary in length and thickness.

This mineral specimen is the same as the other one in composition.

Ⅱ. 表语译作主语

Matter is anything that occupies space and has weight.

Neutrons are of vital important to atoms since they overcome the natural tendency of protons to fly away from one another.

Ⅲ. 主语译作定语

这种译法是与更换主语同时进行的,可参看宾语译作主语的译例。

Ⅳ. 主语译作宾语

对于被动语态作谓语动词、"there"作引导词,以及"find"作"有""在"的句型,英译汉时有时可以把原来的主语改作宾语。

As the match burns, heat and light are given off.

Mountain glaciers are found in all the parts of the world where mountain ranges are high.

There is a tremendous amount or fresh water beneath the surface of the earth.

Ⅴ. 定语译作谓语

英译汉时把原来修饰某一名词的定语译作谓语,或让它与该名词一起构成主谓结构。

The earth was formed from the same kind of materials that makes up them.

Diamonds are characterized by very great hardness.

Ⅵ. 状语译作补语

英语中用作状语来修饰动词和形容词的某些词所表达的意思恰如汉语中补语所表达的意思。

The sun is far bigger than the earth.

The work has been done carefully.

Reading Material

Sedimentary Rocks[②]

Some sedimentary rocks have been formed from igneous rock. The surface layers of igneous rocks have been constantly acted upon by erosive forces, such as changes in temperature, running water, and the blasting effects of fragments hurled by the wind. As a result, some of the igneous rock at the surface has been broken down into fragments. These fragments, swept along by winds, glaciers, streams, and shore currents, have been deposited in lakes or in shallow parts of the sea. They have been pressed together under the weight of later accumulations and have been gradually transformed into sedimentary rocks. Sedimentary rock has also been formed from plant and animal remains and through the evaporation of seawater.

The sedimentary layers of the earth's crust make up most of its surface area. Each layer ranges in thickness from a few centimeters to several meters. In some places there are only a few layers. In others, there are vast accumulations of beds, several kilometers thick.

Generally speaking, the older the bed, the more thoroughly the sedimentary rocks have been cemented. The fragments of young rocks are so loosely held together that they may sometimes be quarried by digging with a spade. The older rocks, however, are generally so solid that they cannot be quarried unless they are drilled, blasted with dynamite, and broken into still smaller fragments with pick or sledgehammer.

---

②FOSTER R J. General geology[M]. Columbus, OH: Charles E. Merrill Publishing Company, 1983.

Lesson Five
Layering or Metamorphic Rocks

【1】 Metamorphic rocks may exhibit layers resembling those of sedimentary rocks. Metamorphic layering, or **foliation**, is due to mechanical and chemical changes in the original rock. Grains, crystal, and fossils are shifted or broken up and strung out in linear series. Parallel rows of plate like minerals, not at all related to the original bedding, may form. Shale is changed by pressure into **slate.** This may later become a rock called **phyllite** and, with continued pressure, a crystalline **foliated** rock known as **schist.** Other kinds of rock, such as **sandy** shale and **granites**, for example, are transformed into a more coarsely foliated crystalline rock called **gneiss.**

【2】 Slate is a dark rock with an invisibly fine grain. It splits readily into thin smooth **slabs.** Phyllite represents a stage intermediate between slate and schist. Its grain is very fine, consisting primarily of mica, but here and there a few larger crystals appear. The foliation is rougher and wavier than that of slate. Schist is visibly crystalline, with **wavy foliation.** It consists such **flakelike** or **tabular foliar** minerals as mica, **chlorite**, **hornblende**, or **talc**, as well as distinct crystals of **quartz** and garnet.

【3】 Gneiss looks irregular and streaky because it has alternating layers of different minerals. Gneisses originate from several different kinds of rock. Each type of gneiss has

foliation n. 叶理,面理
slate adj. 板岩
phyllite n. 千枚岩
foliated adj. 叶片状
schist n. 片岩
sandy adj. 砂质的
granite n. 花岗岩
gneiss n. 片麻岩
slab n. 板片
wavy adj. 波状的,起伏的
flakelike adj. 薄片似的,石片似的
tabular adj. 板状的
foliar adj. 叶的,叶状的
chlorite n. 绿泥石
hornblende n. 角闪石
talc n. 滑石
quartz n. 石英

---

①苏生瑞. 地质专业英语[R]. 唐山:唐山工程技术学院,1986(内部资料).

much the same composition as the mother rock, granite gneiss, for example, is derived from granite.

【4】Other kinds of metamorphic rocks show no foliation at all. Pure sandstones are changed into a more compact mass called quartzite, where pore spaces have been compressed, making a very hard and **durable** rock. **Limestone** is converted by heat and pressure into **marble**, where the carbonate grains become visibly crystalline. Pure marble is white. Most limestone however contains impurities that often react under **metamorphism** to produce new minerals. These often **impart** striking colors or **mottling** marble, making it more attractive as a building stone. Marble is very **plastic** under pressure.

durable *adj.* 持久的,耐久的
limestone *n.* 石灰岩,碳酸钙
marble *n.* 大理岩
metamorphism *n.* 变质作用
impart *v.* 给予,传递
mottling *n.* 斑点构造,成斑作用
plastic *adj.* 塑性的,塑造的

Phrases and Expressions

string out 使成串地展开,连成一列
wavy foliation 波状叶理
alternating layer 互层
pore space 孔隙

Exercises

Ⅰ. Translate the following sentences into English.

1. 尽管成因不同,但变质岩和沉积岩都有类似的层状构造。

2. 变质岩和沉积岩的层状构造的地质意义不同。

3. 岩石经过变质后孔隙被压紧,密度增大。

II. Answer the following questions.

1. Why can metamorphic layering be formed?

2. What is the original rock of gneiss?

3. What is marble changed from?

4. What are the differences between phyllite and slate?

III. Choose an appropriate word from the following box to fill in each of the following blanks. Each word can be used only ONCE. Change the form where necessary.

> fine-grained rocks   mineral composition   chemically active solutions
> stable   retard   metamorphic rocks   endless   weather   shale
> equilibrium

_____ are rocks that have been changed while in the solid state, either in texture or in _____, by any of the following: heat, pressure, directed pressure (stress), shear, or _____. Because any rock of any type or composition can be subjected to any or all of the above agents, the variety of metamorphic rocks is _____.

The changes that occur during metamorphism are the result of an attempt to reestablish _____ with the new conditions to which the rock is now subjected. Again note the similarity to _____. Weathering occurs when rocks that formed deep in the earth are subjected to surface conditions. However, when surface-formed rocks, such as _____, are metamorphosed, the changes that occur proceed in the opposite direction and higher-temperature minerals are formed.

Many factors promote or _____ metamorphic reactions. Large surface area promotes chemical reaction so _____ react faster than coarse-grained rocks. Glass is less _____ than crystalline material and so also reacts faster.

Lesson Five Layering or Metamorphic Rocks

Translation Tips

加词是指在译文中增加某些原文中没有的词,减词是指在译文中省略原文中某些词。加词和减词的目的是确切传达原意,同时力求译文通顺。

Ⅰ. 加词

1. 补出原文省略的部分

Some substances are soluble, while others are net.

The earth is 7.927 miles in diameter at the equator and about 28 miles less in diameter at the poles.

2. 重复原文中的共有成分

Chemical and X-ray analysis reveal the composition of minerals.

Scientists realized that plate tectonics might explain not only the drifting of continents but also the cause of earthquakes.

3. 把原文中隐含的意思表达出来

Most such stars are many lifetimes away.

Often an animal fossil is simply some hard part of an animal's body—its shell, perhaps, or a bone. The soft parts of an animal's body usually decay very fast. The hard parts often last much, much longer.

4. 为了使译文语气连贯,逻辑严密

The earth has been, and is being, affected by a myriad of interacting processes, and it is the task of geology to describe and interpret these processes and their results.

Heat from the sun stirs up the atmosphere, generating winds.

Ⅱ. 减词

有些词类在英语中有，在汉语中没有，如冠词、反身代词。对于这些词，在不影响原意的情况下，英译汉时往往省去不译。另外，英语中还有些句型，如"there"作引导词的句型、用"it"作形式主语或形式宾语的句型，由于这些词本身在句中并无词义，只起语法作用，而汉语中又没有同类结构。因此，英译汉时，将这些词省去不译。这些都不再举例说明。下面将减词的其他情况作简要说明。

1. 省略代词

汉语中一般不重复使用代词，因此英语中很多代词在英译汉时都应省去不译。

Coal is a natural, substance, but it is not classed as a mineral.

Geology, the science of the earth, tells us about the world, what it is made of, its age, its aging processes, its landform, and its abyssal ocean depths.

2. 省略动词

英语句子中动词必不可少，但汉语句子中有时不用动词，因此英译汉时有时可省去动词不译。

In most cases the soil near the surface is merely damp and is not completely saturated with water.

For this reason television signals have a short range.

3. 其他省略情况

其他减词的情况也很多，除了上面讲的，下面再举几例。

In the magnet the atoms are lined up in such a way that their electrons are circling in the same direction.

Already we have learned to split atoms in such a way as to obtain huge amount of energy.

Lesson Five Layering or Metamorphic Rocks

An anticline is, in its simplest form, an arch in which the two <u>sides</u>, or <u>limbs</u>, dip outwards away from one another.

The helium atom has two electrons outside its nucleus, which accounts for <u>the fact that</u> its chemical properties differ from those of hydrogen.

Reading Material

Formation of Metamorphic Rocks[②]

One of the most important factors in metamorphism—the formation of metamorphic rock is pressure. It may be applied by overlying sedimentary beds. It may be caused by magma making its way into surrounding rock layers. It may be due to the mountain-building forces that deform the earth's crust. As pressure is applied, tremendous heat is generated. This quickens the chemical reactions taking place and heightens their effects. The presence of water is another factor in metamorphism. The water may be so scanty that it forms a mere film around the particles. Yet it provides a medium in which rock substances can pass into solution and from which they can condense on the surface of new and growing crystals.

Under extreme heat and pressure, the original rock particles are forced to new arrangements. In some cases, the rock constituents recombine with those in the immediate vicinity and form new minerals, many of which grow with nearly perfect crystal form. Garnet is such a mineral.

[②] FOSTER R J. General geology[M]. Columbus, OH: Charles E. Merrill Publishing Company, 1983.

Lesson Six
Fossils and How They Are Made[①]

【1】Fossils are **traces** of the living things of long ago. All that we know about the plants and animals that lived on the earth before there were any people on the earth has come from fossils. Let us find out how the fossils which tell the stories of the animals of yesterday were made.

【2】An animal fossil may be a whole animal which, when it died, was buried in such away that its body was kept almost as it was when the animal was **alive**. Nearly 50 million years ago, for example, many **insects** were **trapped** in the **sticky resin** which came from the **pine** trees of that time. The resin kept the bodies of the insects from decaying. Some of this resin was buried under the ground. There, it **hardened** into amber. In the amber, the insects trapped in the resin have been kept perfectly.

【3】Often an animal fossil is simply some hard part of animal's body—its **shell**, perhaps, or a bone. The soft part of an animal's body usually **decays** very fast. The hard parts often **last** much, much longer.

【4】Some animal fossils are hard parts of animals that have been **petrified**, or **remade** in stone. If a bone, for example, is buried, water in the ground may take away, particle by particles, the material of which the bone is made and leave some minerals, such as lime, in its place. At last none of the real bone is left. Instead, there is a "bone" of stone—a petrified bone.

trace *n*. 痕迹
alive *adj*. 活着的
insect *n*. 昆虫
trap *v*. 使陷入，圈闭
sticky *adj*. 黏的
resin *n*. 松脂，松香
pine *n*. 松树
harden *v*. 使变硬，变坚固
shell *n*. 贝壳
decay *v*. （使）腐烂
last *v*. 持续，耐久
petrify *v*. （使）石化
remake *v*. 重制

---

①苏生瑞. 地质专业英语[R]. 唐山：唐山工程技术学院，1986（内部资料）.

Lesson Six Fossils and How They Are Made

【5】A fossil may be only a **footprint** of an animal. A **dinosaur**, for example, may have **wandered** along the **shore** of a **pond** one day and left footprints in the mud. The waves may have **washed** sand or mud into the footprints so **gently** that they were not **disturbed**. The layer of mud on which the footprints of the **giant reptile** were made may, as the ages passed, have hardened into solid rock and kept the footprints as they were.

【6】A fossil may be a cast. Have you ever made lead soldiers by pouring lead into a **mold**? If you have, it will be easy for you to understand how a fossil cast may be made. The shell of a **snail** of long ago, let us say, was pressed down into mud. The material in the shell was dissolved by water, but a hole, or mold, was left where the shell had been. Later the holes were filled with **lime** and mud, and the lime and mud hardened into a cast of the shell.

【7】As you probably know, much of the dry land of today was once covered with water. Layers of such rocks as sandstone, and shale—rocks that were made under water—show that it was. Such layers of rock furnish most of the fossils of the living things of past ages.

footprint n. 脚印
dinosaur n. 恐龙
wander v. 徘徊
shore n. 岸
pond n. 池塘
wash v. 洗，冲刷
gently adv. 轻轻地
disturb v. 扰乱，弄乱
giant adj. 巨大的，庞大的
reptile n. 爬虫，爬虫类
mold n. 模子
snail n. 蜗牛
lime n. 石灰

Phrases and Expressions

keep…from   使……免于

Exercises

Ⅰ. Translate the following sentences into English.

1. 在岩石中发现了恐龙化石。

2. 过去的许多动物只留下它们的坚硬部分而成为化石。

3. 泥岩不能逐渐变硬,成为坚硬的岩石。

4. 动物的尸体经过若干年后有可能被埋藏在地下。

II. Answer the following questions.

1. What is a fossil?

2. Why can we sometimes find bodies of insects in amber?

3. What does the word "petrify" mean?

4. Why are most fossils found in sedimentary rocks, such as sandstone, limestone and shale?

III. Choose an appropriate word from the following box to fill in each of the following blanks. Each word can be used only ONCE. Change the form where necessary.

```
shell    uplift    fossil    rapid    preserve    erosion
decay    reconstruction    evidence    incomplete
```

Fossils are any evidence of past life, and this _____ may be preserved in many different ways. When most animals die, the body is _____ destroyed; scavengers consume the body or bacteria cause it to _____. Only a very few escape this destruction, possibly to become _____. Generally, if the body has hard parts, those parts are most likely to be preserved; and if the body is rapidly buried, its _____ is more likely. Burial may also bring on physical and chemical changes that may cause destruction. Later _____ is necessary to expose the fossil and weathering and _____ may destroy it then. Thus fossilization is a relatively rare event, and this is why the record of fossil life is so _____. For these reasons, the most common fossils are the _____ of shallow-water marine animals such as clams. The incompleteness of the fossil record makes _____ of past life very difficult.

Translation Tips

在科技英语文章写作中,定语从句的应用相当广泛,翻译起来有时也比较困难。其原因是:从形式上讲,英语定语从句往往结构复杂,"枝蔓丛生",而汉语则忌用冗长而复杂的定语。

从作用上讲,英语定语从句往往超出单纯修饰的范围,而含有诸如原因、结果、目的、让步等这样一些状语的"意味",而汉语的定语只起修饰或限制的作用。

总之,英语中的定语从句是汉语中没有对应结构的一种较为复杂的语法现象,英译汉时,必须根据具体的情况用不同的方式去处理。下面介绍常用的几种处理方式。

Ⅰ.译成定语("的"字结构)

Rocks and minerals are materials which make up the earth's crust.

The types of data which geologists collect are quite varied and numerous.

The rate at which sediments accumulate varies widely.

Ⅱ.译成平行分句

对于不带有状语意味的非限定性定语从句,一般都可采用分译的办法。分译又可分为前置(定语从句译出后放在主句前)和后置(放在主句后)两种。

Volcanic ash settles quickly to form extensive beds of tuff, many of which contain perfectly preserved fossils.

In addition to these two types of folds, however, we must mention another one, which is not easily classifiable as either anticline or syncline and is referred to as a monocline.

Water itself may, as we have seen, be decomposed to hydrogen and oxygen.

Ⅲ. 译成表示原因、结果、目的、让步等的状语分句

英语的定语从句,无论限制性的或非限制性的,都有可能带有原因、结果、目的、让步等状语的意味。遇到这样的定语从句,应该分别加上"因为""因而""使""为了""以便""让""无论"等词,将原文的定语译为汉语的带有状语意味的分句,以便准确转达原义。

An animal fossil may be a whole animal which, when it died, was buried in such a way that its body was kept almost as it was when the animal was alive.

The sun heats the earth, which makes it possible for plants to grow.

A gas occupies all of any container in which it is placed.

Ⅳ. 译为其他

In determining the age of rocks, several principles can be applied. The principles which are used depend on the type of rock and the results of the earth movements.

Part of the water which falls as rain sinks into the ground.

There is no place on the earth where the days are longer in winter than in summer.

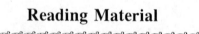

Fossils in the Rocks[②]

The waters of the earth for millions of years have been the home of water plants

---

[②]FOSTER R J. General geology[M]. Columbus, OH: Charles E. Merrill Publishing Company, 1983.

Lesson Six Fossils and How They Are Made

and animals. Some of these plants and animals or their parts have been trapped in the muddy bottom where scavengers were able to eat them or bacteria cause them to decay. Tree trunks, bones, shells, and other solid materials have been washed from the land into the oceans by streams and also trapped in sediments. When the sediments changed to rock, these remains are known as fossils. Fossils are evidence of the forms of life in ancient times.

Although you may see no living thing on a field trip, you can find indirect evidence of the presence of life. The evidence can appear in many forms. You may find the footprint of a deer in the mud. You may find the name of a dement contractor in a concrete sidewalk or a bone that a dog buried after he got tired of gnawing on it. Leaves or twigs may be trapped in the muddy bottom of a puddle. If these subjects were to be preserved in these places for millions of years, they would become fossils.

Geologists find fossils in sedimentary rocks. These fossils are evidence of life that existed in the past. Some of these forms are indirect evidences, that is, the fossils are not the preserved remains of the plant or animal. A footprint or a leaf print only indicates that some animal or plant made them in sediments of long ago. Geologists know that the organism had to exist in order to leave this kind of record. Other fossils are direct evidence because all or part of an organism is preserved. An expert on fossils can identify the kind of organism.

Lesson Seven
Description and Classification of Joints

【1】 **Joint** faces, or even the traces of joints, are rarely seen in their entirety because the outer surfaces of most rock exposures are composed of a number of intersecting and **abutting** joint faces, as well as bedding planes, **schistosities**, faults, eroded surfaces, and man-made **fractures**. For this reason relatively little is known of the complete three-dimensional shapes of joints. Instead, most studies of joints have focused on the more readily observed features, particularly orientation and **spacing**.

【2】 What most readily attracts one's attention of what is visible of joints is their remarkable smoothless and their existence in nearly parallel sets that cut across other sets with no apparent interaction of offset. Joints may be symmetrically oriented with respect to other structures in the **outcrop**, for example, they may be perpendicular to bedding, fold axes, or planar and linear fabrics such as slaty cleavage. Some joints are closely parallel or perpendicular to the surface of the earth. In other cases they are seemingly unrelated to any other structures or surfaces.

【3】 The joints that have been most studied and are best understood are those that in some way systematic or regular in their arrangement; these are called systematic joints. The term joint set is applied to all systematic joints within a region that are parallel to one another and may be distinguished from

joint   *n.* 节理,接头
abutting   *adj.* 邻接的,毗连的,紧靠的
schistosity   *n.* 片理
fracture   *n.* 断裂
spacing   *n.* 间距,间隔
outcrop   *n.* 露头

---

① 苏生瑞. 地质专业英语[R]. 唐山:唐山工程技术学院,1986(内部资料).

other sets of different orientation. Joint sets are said not to affect one another, but rather crosscut without deflection.

【4】 In practice, the recognition of individual joint sets may be aided by plotting a representative sample of measured joint orientations as poles on a spherical projection or by plotting **strikes** of vertical joints on a rose diagram, which is a circular histogram of orientations. Spherical projections and rose diagrams of joints are useful in recognizing potential planes of slip in mining and civil engineering.

【5】 The term joint system, in contrast with joint set, is applied to two or more joint sets that are thought to be generally related—for example, conjugate sets of shear joints, which systematically maintain acute dihedral angles of about 5° to 60° between each other. Another example of systematic joint system is columnar jointing, well known in lava flows, **dikes**, and **sills**, which is an effect of inhomogeneous thermal contraction during cooling of the lava. Systematic joints of the same set or system are sometimes characterized by a distinctive wall-rock alteration, one of the few properties of joints that allow their relative ages to be determined.

【6】 Not all joints are developed in systematic sets and systems; there are also many less regular fractures, which are called nonsystematic joints. Nonsystematic joints usually meet, but do not cross, other joints. As a group, they are less smooth and planar than the systematic joints.

strike　*n.* 走向
conjugate　*adj.* 共轭的
dihedral　*adj.* 二面的
dike　*n.* 岩墙
sill　*n.* 岩床，海底山岩

Phrases and Expressions

bedding plane　层面
eroded surface　侵蚀面
fold axis　褶皱轴
planar fabric　面状组构
linear fabric　线状组构

slaty cleavage　板劈理
systematic joint　系统节理
joint set　节理组
spherical projection　球面投影
vertical joint　垂直节理
rose diagram　玫瑰花图
shear joint　剪节理
conjugate sets of shear joints　共轭剪节理组
acute dihedral angle　锐二面角
columnar jointing　柱状节理
lava flow　熔岩流
wall-rock　围岩
wall-rock alteration　围岩蚀变

Exercises

Ⅰ. Translate the following sentences into English.

1. 露头上很少看到完整的节理面。

2. 我们应把重点放在研究节理的方位和间距上。

3. 有的节理垂直于层面,有的则平行于层面。

4. 工程地质学中常用节理玫瑰花图。

Ⅱ. Answer the following questions.

1. Why are joint face rarely seen in its entirety?

2. What are the best understood joints?

Lesson Seven Description and Classification of Joints

3. Is joint set different from joint system?
_____

4. Are all joints developed in systematic sets?
_____

III. Choose an appropriate word from the following box to fill in each of the following blanks. Each word can be used only ONCE. Change the form where necessary.

```
fold    granite    metamorphose    parallel    sedimentary rock
overlie    disconformity    angular unconformity    deposite    unconformity
```

If sedimentary rocks _____ metamorphic or granitic rocks, the sediments were deposited after the time of metamorphism or intrusion. _____ and metamorphic rocks form deep in the earth, so the erosion that uncovered them required time. In this case, there are no sedimentary rocks representing the time of intrusion or _____ and the time required for erosion. This type of _____ is called a nonconformity. Nonconformities occur where sedimentary rocks are deposited on igneous or metamorphic rocks that formed at some depth in the crust.

Another type of unconformity called a _____ may consist of a change in fossils, representing a short or a long period of time, that occurs between two _____ beds in the sedimentary section. Such an unconformity may be due to erosion of previously _____ beds or nondeposition, and may not be at all conspicuous until fossil are studied.

The third type of unconformity is more obvious and consists of _____ and eroded sedimentary rocks that are overlain by more sedimentary rocks. Such an unconformity is called an _____ because the bedding of the two sequences of _____ is not parallel. In this case, the time of folding and erosion is not represented by sedimentary rocks.

Translation Tips

英语句子中,凡不必、不愿或无从说出动作的发出者时,谓语都要用被动语态。另外,为了强调动作的对象或为了行文需要,也可以用被动语态。因此,英语中被动句式的使用相当广泛,特别是在科技文章中。但在汉语中,即使句中主语是动作的承受者,也常常不必在谓语动词前加上表示被动含义的词。此外,当无须指出动作的发出者时,还可以用无主句形式。因此,被动句的翻译可以采用各种形式。现将被动句的常见译法作如下介绍。

Ⅰ. 保留原来的主语,谓语译为被动式

在谓语前加"被"。

"被"字在汉语中主要用来表示不如意或不期望发生的事,在现代汉语中,它的使用范围似有扩大的趋势,但毕竟是有限的,所以只有一部分英语被动句译为汉语时才在动词前加"被"。

Thus, steel, cement, plaster, and glass, although all are derived from naturally occurring mineral raw materials are not regarded as minerals themselves, since they have been processed by man.

At the edges of the continents, the crust is dragged downward to form the deep trenches of the ocean floors.

用"由""受""为""让""给""加以""予以""为……所"、"是……的"等引出行为主体。

The atmosphere is made up of gases surrounding the solid and liquid parts of the earth.

Much of the dry land of today was once covered with water.

Today the theory of evolution is accepted by almost all scientists.

Ⅱ. 保留原来的主语,谓语译成主动式。

Earthquakes are concentrated in certain areas of the earth.

Knowledge gained from studying earthquakes waves has been applied in various fields.

This idea of evolution is supported by many kinds of evidence.

The forces causing the drift have not been fully explained.

Ⅲ. 译为无主语

Many explanations have been proposed.

Pure oxygen is given patients in certain circumstances.

Ⅳ. 谓语分译

先把原文的被动语态谓语单独译出,放在最前面,再把其他部分译出放在后面。

The outer core is believed to be composed mostly of iron and nickel in a molten state at a very high temperature.

All bodies are known to possess weight.

Reading Material

Joints[2]

Rock at the surface of the earth is cut by a variety of fractures and cracks, into which roots force their way and water seeps. Deeper within the earth these fractures become less and less common, as is sometimes noted in deep mines and quarries. Nevertheless, some cracks still exist at intermediate crustal depths, particularly in plutonic igneous rocks that have cooled substantially since crystallization. Cracks of deep origin are normally heated with vein minerals. Cracks and fractures are a widespread structure feature of the brittle upper part of the crust.

---

[2] FOSTER R J. General geology[M]. Columbus, OH: Charles E. Merrill Publishing Company, 1983.

Any thin natural planar crack that is not a fault, bedding or cleavage and is larger than the grain size of the rock is a joint in the broadest sense of the word. The word is said to have originated with British coal miners who thought the rocks were "joined" along these fractures, just as bricks or building stones are joined together in building up a wall.

Joints are of considerable practical importance. They are a widespread plane of potential slip and therefore must be considered for safety and economics in quarrying, mining, and civil engineering. The orientation and spacing of joints can significantly affect the ease of mining and subsequent handling of coal and some ore. Joints are important to groundwater hydrology and the design of dams because they affect porosity and permeability. Joints are paths for circulation of hydrothermal ore-forming solutions.

Lesson Eight
Recognition of Faults[①]

【1】 The question of which criteria are valid for the recognition of faults is important because faults are largely hidden structures, whose existence generally must be inferred indirectly. Fault-related deformation **renders** most fault zones more **susceptible** to **erosion** than the surrounding country rock; therefore, the existence, position, and orientation of faults is commonly inferred from observations of structural and stratigraphic discontinuity, as well as related physiography. The consideration of criteria for the recognition of recently active faults is especially important because the geological profession is being called upon to make evaluations of potential hazards associated with dams, power plants, and general construction at specific sites in **tectonically** active areas.

【2】 The principal criteria for the recognition and location of faults are ① structural discontinuity, ② **lithologic** discontinuity, ③ fault-zone deformation, often with associated weakened rocks with poor outcropping characteristics, ④ fault-related deformation of the land surface in the case of recent faulting, and ⑤ fault-related **sediments** and **sedimentation** patterns in the case of syndepositional faulting. Several examples of the application of these criteria and related phenomena are given in the following paragraphs.

【3】 Many faults are inferred to lie along mapped lithologic or structural discontinuities. For example, the great **Sarv thrust**

render　v. 反映,表达
susceptible　adj. 受……影响的
erosion　n. 侵蚀
tectonically　adv. 构造上地
lithologic　adj. 岩性(学)的
sediment　n. 沉积物,沉淀物
sedimentation　n. 沉积物形成作用,沉积作用
syndepositional　adj. 同沉积(作用)的
Sarv　n. 萨夫(地名)
thrust　n. 逆断层,逆冲断层

---

① 苏生瑞. 地质专业英语[R]. 唐山:唐山工程技术学院,1986(内部资料).

sheet of the **Scandinavian Caledonian** mountain belt is characterized by a north-striking **diabase**-dike swarm that **intrudes** the Upper Precambrian **quartzite**; the dikes are missing in the underlying thrust sheets. The lack of correlative dikes in the lower plate of the Sarv thrust sheet exists for more than 100 km across strike, suggesting a net slip of at least this value. Another example involving dikes is the **Garlock** fault in eastern California, for which a 60 to 70 km offset of apparently correlative dike swarms and granitic country rock suggests a large left-lateral strike slip. **Truncated** dike swarms in these two examples represent a combined lithologic and structural discontinuity.

【4】Stratigraphic repetition, omission, and discontinuity are among the most important evidence of faulting, although **facies** changes and **interfingering** sediments are occasionally confused for faults. Much of the evidence for thrust faulting within the **Appalachian** Valley-and-Ridge province and the **Alberta** Fold-and-thrust belt comes from the stratigraphic repetitions observed in the surface geology; the actual faults are generally not exposed. As you travel across eastern **Tennessee**, the same **Cambrian** and **Ordovician** section is repeated six times, always **dipping** eastward.

Scandinavian　n. 斯堪的纳维亚(地名)
Caledonian　n. adj. 加里东期(的)
diabase　n. 辉绿岩
intrude　v. 侵入
quartzite　n. 石英岩
Garlock　n. 加洛克(地名)
truncate　v. 截去……的顶端，截断，截短
facies　n. 相
interfingering　adj. 相应穿插的，指状交叉的
Appalachian　n. 阿帕拉契亚(地名)
Alberta　n. 亚伯达(地名)
Tennessee　n. 田纳西(地名)
Cambrian　n. adj. 寒武纪(的)，寒武系(的)
Ordovician　n. adj. 奥陶纪(的)，奥陶系(的)
dip　n. 倾向

Phrases and Expressions

country rock　围岩
stratigraphic discontinuity　地层间断
call upon　号召，指派
syndepositional faulting　同沉积断层
Upper Precambrian　上前寒武统(的)
thrust sheet　逆掩盘
net slip　总滑距，净滑距
left-lateral　左行的
stratigraphic repetition　地层重复

stratigraphic omission 地层缺失
fold-and-thrust belt 褶皱-逆冲带
actual fault 实际断层

Exercises

I. Translate the following sentences into English.

1. 识别节理和断层的标准有一定的差别。

2. 若地层间断,则可能有断层存在。

3. 不论是节理还是断层,对大坝都有潜在危害。

4. 这里奥陶纪地层缺失。

II. Answer the following questions.

1. Where is the existence, position, and orientation of faults inferred from?

2. Why are the criteria for recognition of active faults especially important?

3. What is the most important evidence for faulting?

4. Where do many faults lie?

5. Give the principal criteria for the recognition are location of faults?

Ⅲ. Choose an appropriate word from the following box to fill in each of the following blanks. Each word can be used only ONCE. Change the form where necessary.

```
erosion    curve    undercut    valley-side    cut
wide    meander    headward erosion    deflect    width
```

Erosion tends to deepen, lengthen, and _____ a river valley. The ways that a river can deepen its channel were just discussed. At the head of a stream any downcutting lengthens the valley. This process is called _____ and is the way that streams eat into the land masses. A stream valley is widened by two methods. The _____ processes, such as creep and land sliding, widen the valley. Lateral _____ by the stream itself is the other process.

Lateral cutting is most pronounced on the outsides of curves where the valley sides may be _____. Thus a curving stream widens its valley, and straight stretches of screams longer than ten times the _____ are not common. Clearly, any irregularity on a scream bottom or bank can _____ the flow and cause a curve to develop. Once formed, a curve tends to migrate, thus widening the valley. Many rivers have distinctive symmetrical curves called _____. The meander-forming curve is the one that distributes the river's loss of energy most uniformly. Any other curve would tend to concentrate _____ and deposition at some point on the _____.

Translation Tips

英汉两种语言表达否定的方式颇有差异。下面将英汉翻译中否定表达式的译法应着重注意的地方略加举例说明。

Ⅰ. 否定位置的转移

否定词在句中究竟否定哪个部分,对全句的理解与翻译较为重要。从英译汉的角度来看,英语中的否定谓语,译成汉语时却转换为否定状语;英语中的否定宾语,译为汉语时却应是否定谓语等。这些在翻译时是必须十分注意的。

Perhaps you do not think of glass as an elastic material.

He does not speak English correctly.

Some of these microscopic creatures have no particular shape.

The whole subject is so obscure that I have succeeded in throwing hardly any light on it.

The motor did not stop running because the fuel was finished.

Ⅱ. 部分否定与全部否定

1. 部分否定的表达方式及译法

A. not all … 句式

Not all strike-slip faults are vertical, nor is the slip always 100 percent horizontal.

Not every minute difference is noticed.

Not both of them serve the purpose.

B. all/every … not … 句式

All synclines do not increase the return energy.

Every subject is not treated in the same way.

C. … not … all 句式

He does not like both.

This theory of convection currents within the earth is not yet accepted by all geologists.

2. 全部否定的表达式及译法

A. none, no, neither, never＋肯定式谓语

None of these strike-slip faults are vertical.

Neither of the substances dissolves in water.

B. 肯定式谓语＋none, neither, nothing

A proton has a positive charge and a electron a negative charge, but a neutron has neither.

I saw nothing.

The straight line passes through none of the points.

Ⅲ. 把否定形式译为肯定形式

有些英语句子形式是否定的,内容却是肯定的,英译汉时可以把这类句子译为肯定式。

The leaves, which are green now, will not turn red until the weather becomes frosty in late autumn.

Metals do not melt until heated to a definite temperature.

It is not until meteors strike the earth's atmosphere that they can be seen.

It was not until the sixteenth century that a constructive effort toward solving the ocean's mysteries was begun.

It not in frequently happens that what were thought to be correct statements of natural laws turn out to be wrong, or at least subject to modification.

There is nothing unexpected about it.

Reading Material

Faults[2]

Suppose that stresses develop within the lithosphere that tend to move one block of the crust northward and the neighboring block toward the south; or to depress one block and elevate the adjacent one. A fracture may then form between the two blocks, along which movement will take place so as to relieve the stress. Such fractures, termed faults, are surfaces along which the rocks on one side are displaced with respect to those on the other. Tiny faults in hand specimens may show displacements of a fraction of an inch; at the opposite extreme are faults along which the rocks on opposite sides are offset for many miles.

Fault surfaces may occupy any position from vertical to horizontal. Faults may cut any kind of rock and displace any type of geologic structure. Movement may be vertical, horizontal, or a combination of the two. Thus a full classification of fault displacement is a highly involved exercise in solid geometry. For descriptive purposes, however, it is enough to distinguish three major kinds of faults.

A strike-slip fault is one on which the rocks on opposite sides are displaced horizontally (Figure 8-1). Because the displacement or "slip" is parallel with the strike of the fault, the term strike-slip fault is appropriate. Not all strike-slip faults are vertical, nor is the slip always 100 percent horizontal; but these are the dominant attributes of strike-slip faults and separate them from the other types.

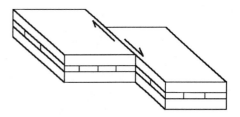

Figure 8-1 A strike-slip fault

A normal fault is quite different from the one just described. For one thing, the relative movement has been up-and-down rather than horizontal; for another, the fault surface is not vertical but inclined(Figure 8-2). We call the rocks that lie above an

---

[2] FOSTER R J. General geology[M]. Columbus, OH: Charles E. Merrill Publishing Company, 1983.

inclined fault the hanging-wall block and those that lie below it the footwall block. We use these terms because they make it easy to describe the movement on such faults. In the figure, it is clear that the hanging-wall block has moved down relative to the footwall block.

A reverse fault is one on which the hanging-wall block has moved up relative to the footwall block(Figure 8-3). Because the ends of the block are closer together after faulting than before, we conclude that reverse faults are produced by compression of crustal rocks.

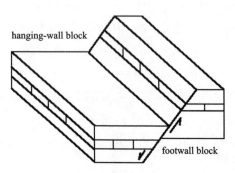

Figure 8-2　A normal fault

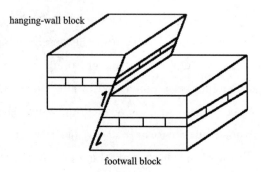

Figure 8-3　A reverse fault

Lesson Nine Parallel Folds

【1】 Most **folds** are **cylindrical**, that is, they are composed largely of cylindrical or nearly cylindrical (highly **elliptic**) points with the directions of minimum principal **curvature** (fold axes) parallel or approximately parallel. The variety of shapes of cylindrical folds can therefore be considered by simply viewing them in **profile** section—that is, by looking at the shapes of lines of principal maximum curvature. In this section we consider the shapes of single surfaces and layers; in a later section we consider how adjacent layers stack together to form a folded volume of rock.

【2】 There is an important division of cylindrical single layer folds into parallel and nonparallel types. Parallel folds maintain approximately constant layer thickness around the fold measured perpendicular to bedding; nearby bedding surfaces are parallel. The essence of parallel folding from a mechanical point of view is that original layer thickness is preserved; the layers have undergone little **stretching**. Some folds that are not parallel because of originally variable stratigraphic thickness are nevertheless parallel in a mechanical sense because of preservation of original layer thickness and lack of important stretching. In contrast, nonparallel folds have a variable-layer thickness measured perpendicular to bedding. The variable thickness is generally a result of deformation during folding, commonly with thickening in the **hinge** and

fold　*n.* 褶皱
cylindrical　*adj.* 圆柱形的
elliptic　*adj.* 椭圆形的
curvature　*n.* 曲率
profile　*n.* 剖面(图), 廓线, 轮廓
stretching　*n.* 伸展, 伸长
hinge　*n.* 枢纽

---

② 苏生瑞. 地质专业英语[R]. 唐山: 唐山工程技术学院, 1986(内部资料).

thinning on the fold **limbs**.

【3】Parallel folds must form largely through layer-parallel slip with little other fold-related distortion in order to display constant layer thickness. Most of the shortening of a single-layer parallel fold is therefore accomplished largely by deflection of the layer; only a small amount of shortening is accomplished through distortion of rock within the layer, largely in the more strongly curved hinge areas.

【4】Some parallel folds are nearly continuously and smoothly curved; the distortion is broadly distributed through the layer. In contrast, other parallel folds are sharply angular, with most curvature and distortion concentrated in a narrow hinge zone. For these reasons, it is convenient to divide parallel folds into curved parallel folds and angular parallel folds(Figure 9-1); a complete graduation can be found between the two types.

limb　*n.* 翼

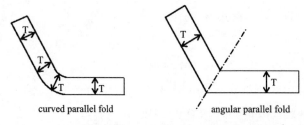

Figure 9-1　Two types folds

* T: the layer thickness, refers to the vertical distance between the upper and lower boundaries of a single, relatively homogeneous layer of sedimentary rock or stratum. T 为岩层厚度,是指单个相对均质的沉积岩层或地层从上边界到下边界之间的垂直距离。

Phrases and Expressions

principal curvature　主曲率
cylindrical fold　圆柱形褶皱
profile section　剖面(图)
single-layer　单层的

Lesson Nine Parallel Folds

Exercises

Ⅰ. Translate the following sentences into English.

1. 在剖面上常见的褶皱是非平行褶皱。

2. 断层可分为正断层、逆断层和平移断层。

3. 岩层在褶皱的过程中有的变厚,有的变薄。

4. 褶皱的要素包括枢纽、轴面和翼等。

Ⅱ. Answer the following questions.

1. What are most folds composed of?

2. How can the variety of shapes of cylindrical folds be considered?

3. Why are some folds not parallel?

4. How does parallel fold from?

Ⅲ. Choose an appropriate word from the following box to fill in each of the following blanks. Each word can be used only ONCE. Change the form where necessary.

> nature   deformation   scale   press   occupy
> fault   how   faulting   feature   deform

The continents display structures on many _____. Bent, broken, and otherwise _____ rocks are seen in small and large exposures. Study of these deformed rocks reveals much information on the kinds of forces that caused the _____.

The occurrence of these deformed rocks in distinct belts shows larger, more fundamental structures that _____ many thousands of square miles. These larger structures are best developed in mountain belts. Further study reveals many other unusual _____ of these large belts of deformed rocks and suggests that they may show _____ continents form. Thus, the article progresses from the small, easily, studied features of deformed rocks to much large, more conjectural structures.

When rocks are subjected to forces, they may either break or be deformed. Abundant evidence of both kinds of response is common as _____ and folds, respectively. Whether a rock breaks or is folded depends on its _____, which depends on the type of rock and the prevailing temperature and _____. Thus ductile rocks tend to be faulted. Folding is favored at depth, and _____ is more common near the surface.

Translation Tips

翻译用 it 作形式主语的句型时,把 it 作形式主语的主句译成一个独立短语或独立分句。这样,作为真正主语的从句本身就可以分离出来,从而恢复它应有的主体地位。

It is clear that continued compression on a rock layer may result in a fault.

___

It is common knowledge that the earth rotates round the sun.

___

It is generally believed that oil is derived from marine plant and animal life.

___

It is proved that heat is a form of energy.

___

It seems to me that hardiness is the chief essential for success.

___

Folds[2]

Folded rock layers are among the most aesthetically deformation structures. Furthermore, they are important structures because a great deal of natural deformation is accomplished by the folding of rock layers. Folds range in scale from microscopic in some fine-grained metamorphic rocks to hundreds of kilometers across in epeirogenic warps.

The important physical mechanisms of folding are not limited to rocks, because the mechanisms depend primarily on the layered nature of the material and not on the fact that it is rock. Analogous folded structures are produced in layered materials composed of paper, wood, metal, rubber, plastic, or viscous liquids. Layered materials squirm when you squeeze them; the layers buckle to the side rather than sustaining the brunt of the compression. For example, the layers of your belly will buckle into a retain of folds as you slouch deeper in your chair. Much folding in rocks is accomplished by buckling in an analogous way (Figure 9-2), but of course the deformation mechanisms of rock are different.

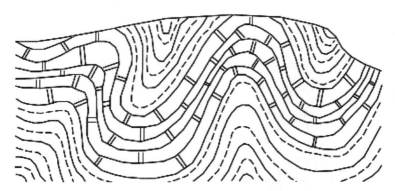

Figure 9-2 The bending and deformation of rock layers

---

[2] FOSTER R J. General geology[M]. Columbus, OH: Charles E. Merrill Publishing Company, 1983.

Lesson Ten
Ground Water

【1】 All water within the ground occurs in cracks and **cavities** in rocks or in the pore spaces between individual mineral grains of soil or rock. If one **penetrates** deeply enough into the ground, he will come to a level below which all space not occupied by mineral matter is filled with water. Beneath a much greater depth, which varies depending on rock type, the amount of water decreases as one penetrates deeper in the earth, until the rocks are dry.

【2】 The **loose** covering of **weathered debris** above **massive bedrock** constitutes the soil. At some depth within the ground, which varies at different localities and with **rainfall** at a single locality, occurs the **uppermost** level of **saturation** with water. This surface marking the top of the **saturated** zone is called the water table. In **swamps** the water table is near the ground surface, but in deserts it lies far down in the bedrock. During **wet** weather the water table approaches the ground surface; however, the position of the water table is lowered at times of **drought**. Why? The term zone of aeration is applied to the region above the water table. Here fractures and spaces between mineral grains contain air as well as some **moisture**, known as vadose water. The zone of aeration is the site of most effective chemical **weathering**, because there mineral grains are in direct contact with oxygen, carbon dioxide, and water. Water in the zone of saturation beneath

cavity *n.* 洞穴
penetrate *v.* 穿过,渗透,钻入
loose *adj.* 松的,松散的
weather *v.* (使)风化
debris *n.* 碎片,岩屑
massive *adj.* 大块的
bedrock *n.* 基岩
rainfall *n.* 降水量,一场降雨
uppermost *adj.* 最上的
saturation *n.* 饱和,饱和作用
saturate *v.* 使饱和
swamp *n.* 沼泽,沼泽地
wet *adj.* 潮湿的
drought *n.* 干旱季节,旱灾
aeration *n.* 通风,通气
moisture *n.* 潮湿,湿气,湿度
vadose *adj.* 渗流的
weathering *n.* 风化作用

---
① 苏生瑞. 地质专业英语[R]. 唐山:唐山工程技术学院,1986(内部资料).

the water table is called ground water.

【3】Water accumulates in the ground from three sources. Nearly all of it originates from the atmosphere as **precipitation**. Secondly, magma contains some water dissolved in the silicate melt, and part of this may escape from the magma chamber to become ground water. Finally, sedimentary rocks formed in the ocean contain sea water trapped in space between mineral grains, and some of this water becomes mixed with other moisture in the ground.

【4】Part of the precipitation that strikes the ground runs off the surface directly into **streams**. Most of the moisture that soaks into the soil returns to the air by evaporation or by **transpiration** of plants. Growing plants require about 40 gal of water to produce 1 lb of plants matter. Much water returns to the ground surface in a rapid flow or as a gradual **oozing** of moisture from the ground. Some is removed from man-made wells. A small percentage of ground moisture comes into the formation of new minerals by chemical weathering, especially the development of clays from other silicate minerals and the formation of **limonite** from other iron-bearing minerals. The remaining moisture accumulates in the earth as ground water.

precipitation　n. 沉淀,降雨量
silicate　n. 硅酸盐
stream　n. 小河,川,溪流
soak　v. 浸,渗透
transpiration　n. 蒸发,蒸腾作用
oozing　v. 渗出,使(液体)缓缓流出
limonite　n. 褐铁矿

Phrases and Expressions

water table　潜水面
zone of aeration　包气带
vadose water　循环水,渗流水
silicate melt　硅酸盐熔融体
magma chamber　岩浆层
soak into　渗入
iron-bearing　含铁的

Exercises

Ⅰ. Translate the following sentences into English.

1. 当雨降落到地上时,大量的水会渗入地下。

2. 在沙漠地区,潜水面位于基岩深处。

3. 潜水面下面的饱和带中的水为地下水。

4. 水库修成之前,在这里没有足够的水供给农田。

Ⅱ. Answer the following questions.

1. Where does all water within the ground occur?

2. Where can we find the saturated zone?

3. What is water table?

4. What is called ground water?

Ⅲ. Choose an appropriate word from the following box to fill in each of the following blanks. Each word can be used only ONCE. Change the form where necessary.

| water table | below | topography | perch | shale |
| control | permeable | sandstone | spring | porosity |

The surface below which rocks are saturated with water is called the _____. Some water is retained above the water table by the surface tension of water. In general, the water table is a reflection of the surface _____, but is more subdued, that is, has less relief, than the surface topography. Lakes and swamps are areas where the land surface is either _____ or at the water table. _____ occur where the water table is exposed as on a valley side. The position of the water table changes seasonally, which explains why some springs are dry in summer. Springs may form where the water table is _____. This situation results where the downward percolation of rainwater is stopped by a relatively impermeable rock, such as a _____. Springs of this type may flow only during the wet season.

The movement of ground water is _____ by the physical properties of the rocks. The amount of water that can be stored in the rocks is determined by the _____, the amount of pore, or open, space. The availability of the water is determined by the interconnections of the pore space, which is, of course, the _____. The most common reservoir rock is _____, although fractured granite or limestone, as well as many other rock types, can serve equally well.

Translation Tips

英语长句一般都有结构复杂,层次重叠的特点,所以长句翻译的困难在于如何重新安排这些层次,从而使译文结构清晰,语义明确,语言流畅。

翻译长句时,需在准确地理解全句意思的基础上,厘清各部分之间的相互关系,然后根据汉语的表达习惯,予以重新安排,而不拘泥于原文的结构形式。下面介绍长句的几种译法。

Ⅰ. 译成带长修饰语的紧长型长句

长句中有一连串的修饰语、从属句或其他成分,它们紧密相连,集中地表达一个完整的概念,翻译时为了保持概念的完整性,句子虽长,一般也要译成中间没有停顿或停顿很少的带长修饰语的句子。

The amount of energy is the amount of positive work the body can do in changing from the condition it is in to some other condition.

The mineral that make up the great bulk of a rock and hence determine both its gross chemical properties and its classification are called the essential minerals.

Ⅱ. 译成舒缓型长句

如果原文句子虽长,但其修饰成分、从属句所表达的概念具有相对的独立性,一般可拆开来译,译成由若干短语或分句构成的舒缓型长句。为了使译文前后语气连贯,逻辑严密,有时可加必要的词语。

1. 顺序译法

如果原文的叙述层次与汉语习惯大体相同,则可按原文的上下层次、顺序翻译。

In fact, the generally accepted theory now is that the production of rain depends upon the presence of dust particles in the air, which serve as nuclei about which the drops may form.

The history of geology is full of plausible generalizations tested and discarded; on the other hand, seemingly half-baked fancies have proven, under close examination and repeated test, to be probable after all—for instance, the theory that the existing continents were once parts of larger masses that split and drifted apart.

---

2. 逆序译法

有些英语长句的叙述层次与汉语习惯相反,若按原文的顺序翻译,则使译文层次不清、概念模糊。因此,必须按汉语的习惯,将叙述层次予以适当调整,有时则须由下而上,逆序翻译。

In reality, the lines of division between sciences are becoming blurred, and science is again approaching the "unity" that it had two centuries ago—although the accumulated knowledge is enormously greater now, and no one person can hope to comprehend more than a fraction of it.

Most of what we know about the earth below the limited depth to which boreholes or mine shafts have penetrated has come from geophysical observations.

Ⅲ. 断句分译法

汉语习惯用短句,即使在表达比较复杂的概念时,出常用一连串的短名,逐层交待,层层铺开,使词句精炼,条理清晰。所以当英语长句中从属句或侵蚀语所表达的内容,相互关系并不密切,具有一定的独立性时,常将长句译成若干短句,然后,按照逻辑顺序,予以重新安排,有的顺译,有的逆序翻译。

Many sedimentary rocks contain such features as ripple marks, exactly like those we can see forming in sediments at the bottom of present-day lake or shallow seas.

Reading Material

The Water Table[②]

You have learned that when rain falls upon the soils, some of it soaks down into the pores, some runs off, some of it may remain close to the surface and return to the air by evaporation, some may be absorbed by the roots of plants and returned to the air by the leaves. The water that remains in the soil seeps downward until all of the open spaces between the mineral grains are filled with water. The water soaks downward until it cannot pass hard-packed layer of soil through which it cannot pass easily.

The soil through which the water passed while seeping down to the hard-packed layer remains damp but not soaked with water. This damp soil with air spaces between the soil grains is the rooting area for plants. When the water fills and saturates the pore spaces of the soil, the ground water geologist (hydrologist) calls the top of this saturated zone the water table. Ground water, therefore, is contained in the pore spaces below the water table.

Water moves downward, but it also moves sideways. Why? You have learned that gravity causes all liquids to flow downward, spread out, and form a flat surface, unless they are acted upon by another force. Capillary action exerts a force upward against the downward pull of gravity.

If water move freely through soil on the surface, the water level in the soil would develop a uniform flat surface, as it does on the lakes. This is not what happens. The force of capillarity upward tends to hold back the movements of water downward through soil. The water table is an average between a uniformly flat level and the shape of the land surface. Water in the saturated zone of the soil flows from the high areas to the low areas.

---

②FOSTER R J. General geology[M]. Columbus, OH: Charles E. Merrill Publishing Company, 1983.

The water table moves up or down in relation to the amount of rainfall. During periods of heavy rainfall, the water table may rise so that its surface is above the ground surface in low spots. Water seeping out of the ground at any of these low spots is called a water table spring. This water will collect on the surface to form a swamp, a pond, or a lake; depending on how much higher the water table is than the ground surface. During periods of dry weather, the water table may fall so much that some springs no longer flow. Then the level of the pond or lake falls, and the pond or lake may even dry up.

Lesson Eleven
Soil Composition[①]

【1】 Engineers classify the earth materials into two broad categories: rock and soil. Although both materials play an important role in foundation engineering, most **foundations** are supported by soil. In addition, foundations on rock are often designed much more conservatively because of the rock's greater **strength**, whereas economics prevents over conservatism when building foundations on soil. Therefore, it is especially important for the foundation engineer to be familiar with soil mechanics.

【2】 One of the fundamental differences between soil and most other engineering materials is that it is a particulate material. This means that it is an assemblage of individual particles rather than being a continuum (a continuous solid mass). The engineering properties of soil, such as strength and **compressibility**, are dictated primarily by the arrangement of these particles and the interactions between them, rather than by their internal properties.

【3】 Another important characteristic that differentiates soil from most other materials is that it can contain all three phases of matter **simultaneously**. The solid portion (the particles) includes one or more of the following materials: ①Rock fragments such as granite, limestone, and **basalt**; ②Rock minerals such as quartz, **feldspar**, **mica**, and **gypsum**; ③Clay minerals such as **kaolinite**, **smectite**, and **illite**; ④Organic matter

foundation　n. 基础, 地基
strength　n. 强度
compressibility　n. 压缩性, 压缩系数, 压缩率
simultaneously　adv. 同时地
basalt　n. 玄武岩
feldspar　n. 长石
mica　n. 云母
gypsum　n. 石膏
kaolinite　n. 高岭石
smectite　n. 蒙脱石
illite　n. 伊利石

---

①DE FREITAS M. Engineering geology: principles and practice[M]. Berlin: Springer, 2006.

such as decomposed plant materials; ⑤ Cementing agents such as calcium carbonate; ⑥**Miscellaneous** materials such as man-made debris.

【4】Liquids and gasses fill the voids between the solid particles. The liquid component is usually water, but it also could contain various chemicals in solution. The latter could come from natural sources, such as **calcite** leached from limestone, or artificial sources such as **gasoline** from leaking **tanks** or pipes. Likewise, the gas component is usually air, but also could consist of other materials, such as **methane**. For simplicity, we will refer to these components as "water" and "air".

【5】A special exception to this three-phase structure is the case of a saturated soil at a temperature below the freezing point of water. These frozen soils are essentially a completely solid material and require special analysis and design techniques.

miscellaneous　*adj.* 混杂的，各种各样的；其他的
calcite　*n.* 方解石
gasoline　*n.* 汽油
tanks　*n.* 贮水池，大容器
methane　*n.* 甲烷，沼气

Phrases and Expressions

foundation engineering　基础工程，地基工程
engineering property　工程特性
cementing agent　胶结物
calcium carbonate　碳酸钙
three-phase structure　三相结构

Exercises

Ⅰ. Translate the following sentences into English.

1. 空气可以视为无质量的。

2. 我们研究描述土集合体的各个组成部分之间的各种关系。

Lesson Eleven Soil Composition

3. 土粒的密度是土粒的体积除它的质量。

4. 如果饱和度已知,就可以确定土的三相组成的比例值。

Ⅱ. Answer the following questions in English according to textbook.

1. What is the difference between moisture content and degrees of saturation?

2. By two important characteristics soil can be differentiated from other materials. What are they?

3. The engineering properties of soil are affected by many factors, example at least three of the factors.

Ⅲ. Choose an appropriate word from the following box to fill in each of the following blanks. Each word can be used only ONCE. Change the form where necessary.

```
disperse    allow    have    stable    fine
net product    silt    hold    quantitate    corner
```

The soil structure refers to the geometric configuration of the particles in a soil aggregate and _____ a profound effect on the physical properties of the soil. Unfortunately, no satisfactory _____ measure has yet been devised to describe the structure.

The structure of natural soils is the _____ of the interaction between the forces of sedimentation, surface forces of the soil particles, and subsequent geologic forces, if particles of sand are _____ to settle from a suspension in water, the particles tend to take up _____ positions to form a single-grained structure.

Very loose sand or _____ may have a honeycomb structure. If the _____ particles consist of clay minerals, the surface forces play an important part. If strong attractive force exist between the edge or _____ and the face of clay plates, a flocculent structure develops. Otherwise, the clay plates may occupy nearly parallel positions as they settle from suspension. This is called a _____ structure. Soils with flocculent and honeycomb structures have large voids between solid particles and are _____ together by surface forces at the contact points. Such structures are generally not very stable.

Reading Material

Soil Classification[②]

Standardized systems of classifying soil are very important, and many such systems have been developed. A proper classification reveals much useful information to a foundation engineer.

Unified Soil Classification System: the most common soil classification system for foundation engineering problems is **the Unified Soil Classification System** (USCS) ASTM D2487. It was proposed by Casegrande and developed by a group of engineers in America. It has been accepted internationally today. The system is also based on the grain size distribution found within soil mass and recognizes three main groups: coarse soil, fine soil and organic soil.

This system assigns a two or four-letter group symbol to the soil, along with standardized descriptions called the group name. Several potential group names are associated with each group symbol. For example, a certain soil might be classified as "SC-Clayey sand", where SC is the group symbol and clayey sand is the group name.

The first letter of the group symbol tells the general type of soil:

G=gravel   S=sand   M=silt   C=clay   O=Organic

The second letter is a supplementary description:

W=well-graded   P=poorly-graded   M=silty

C=clayey   L=low plasticity   H=high plasticity

A special group symbol, Pt, is assigned to peat, which is a highly organic soil that is generally unsuitable for supporting structural foundations.

Chinese System of Soil Classification: according to code for design of building foundation (GB 50007—2002), soil is basically divided into 5 groups including soil of crushed rock, sand soil, silt soil, clay soil and artificial soil.

---

②DE FREITAS M. Engineering geology: principles and practice[M]. Berlin: Springer, 2006.

Lesson Twelve Writing

The Characteristics of Scientific and Technological English Writing

Ⅰ. 正规写作

The writing of scientific and technologic paper should be done in a formal way. 即句子要完整,语法要正确(correct grammar),用词尽量要正规(formal),一般词不用缩写,数字开头要全拼。

Ⅱ. 强调简洁、特性和客观性(Emphasizes its clarity, specificity & objectives)

由于强调客观性,科技英语写作中被动语态(passive voice)较其他在商业、文学等英语写作中用得多。另外,科技英语写作中第一人称用得少,主要用于阐明客观事实和原理,如用人称,常常用"one"泛指。

Ⅲ. 中等写作文体(Middle Style)

科技文章写作通常采用中等文体,而不是高等文体或低等文体。因为科技文章写作重点在于内容而非文章本身,所以不必使用过多华丽的辞藻,要求文章能被读者接受、理解和传播。

High style(高等文体)一般辞藻华丽或使用不常见词汇,句子结构复杂、含义隐晦,这种风格常用于外交和文学领域;

Middle Style(中等文体)是世界大多数科技文章写作所采用的风格。

Low Style(低等文体)多使用口语化的措辞,这种文体不适用于正规写作。

Ⅳ. 尽量避免头重脚轻,以使读者易于理解你表达的意思

科技论文写作的特点之一是结构上的平衡,即尽量避免"头重脚轻"的现象。在开头部分,应避免过多的细节和复杂的理论介绍,即避免"头重"。应尽量在开头部分提供

清晰的背景信息、研究目的和重要性,以便读者能够迅速地理解研究的背景和意义。所谓的"脚轻",指的是论文的结尾部分未能有效总结研究的主要发现、结论和对未来工作的展望。如果论文在结尾处未能给出强有力的总结,读者可能会感到论文缺乏深度和完整性。因此,作者应该在结尾部分明确总结研究的关键点,强调研究的贡献,并提出可能的后续研究方向。

The Skills and Methods for English Abstract Writing

Ⅰ. 英文摘要的写作要求

撰写英文摘要时,要把表达的内容先吃透,再动笔。因为中文和英文不是一种文字,最好不要根据中文摘要逐字翻译。英文的特点是讲究逻辑性,中文的特点是简洁。中文某些表达直译成英文会出现错误,或引起混乱。

摘要是一篇科学文章中最重要的部分,它被阅读的次数可能为文章全文被阅读次数的 500 倍,因此一定要有实质性的内容,不能只是文章中大标题的重复或空洞的阐述,如"… is discussed""… is described"。摘要是科技论文内容和结论的扼要总结,应提供有价值的信息,使同行读者读后决定是否要读全文,使相关学科的读者能获得有用的或有启发性的信息。

摘要写作要注意以下几个方面:①摘要的首句不应是文章标题的重复,应阐述本研究的目的和意义;②摘要应表明新观察到的事实,实验或论点的结论,或者任何新理论、新方法、新技术等的核心部分;③摘要应给出新的化合物、新矿物的名称、新的化石属种名等,以及新的数字资料,如物理化学条件等;④摘要在给出实验结果时,要指出所用方法,如为新方法,应阐明其基本原理,适用范围和准确程度。

另外,要使用完整句,而不是罗列一些标题。文中术语要规范,不能随意创造新名词。应避免使用任何缩约词(contraction)。不应在摘要中摘引他人文章内容。

Ⅱ. 优秀摘要撰写要求

1. 摘要的一般功能

概括主要观点:摘要应提供文献、研究报告或文章的核心内容,概括其主要观点和发现。

决定信息价值:帮助读者快速判断文献或文章是否与自己的研究或兴趣相关,从而决定是否需要深入阅读全文。

扩大传播范围:摘要便于在数据库、期刊或其他媒体上进行快速传播,增加文献的可见度和引用率。

2. 摘要的写作要求

摘要写作时要求有整合性(integrated)、简洁性(concise)、一致性(consistent)和集中性(concentrated)。

3. 摘要的语言特点(Linguistic Features of Abstract)

有限长度(limited length)：摘要是论文的微缩版，其字数有严格的限制。对于较长的论文或报告，通常300个词是一个合理的上限，不超过500个词；对于短篇文章，50~100个词可能就足够。一般而言，摘要的长度是论文长度的3%~5%，一般不超过页面篇幅的2/3。

主要部分(major parts)：从语言结构的维度来看，一个相对完整的摘要通常由以下3个主要部分组成：topic sentence(主题句)、supporting sentences(支撑句)和concluding sentence(s)(结论句)。

* **Topic Sentence(主题句)**

摘要中的第一句话通常被称为"主题句"。通过回答"是什么"的问题，主题句总是直接指向主题或问题，并指出论文的主要目标。主题句常用句式如下。

- The purpose of this paper is …
- The primary goal of this research is …
- The intention of this paper is to survey …
- The overall objective of this study is …
- In this paper, we aim at …
- Our goal has been to provide …
- The chief aim of the present work is to investigate the features of …
- The authors are now initiating some experimental investigation to establish …
- The work presented in this paper focuses on several aspects of the following …
- The problem we have outlined deals largely with the study of …
- With his many years' research, the author's endeavor is to explain why …
- The primary object of this fundamental research will be to reveal the cause of …
- The main objective of our investigation has been to obtain some knowledge of …
- With recent research, the author intends to outline the framework of …
- The author attempted the set of experiments with a view to demonstrating certain phenomena …
- The experiment being made by our research group is aimed at obtaining the result of …
- Experiments on … were made in order to measure the amount of …

• The emphasis of this study lies in …

* **Supporting Sentences**(支撑句)

主题句通常后面会跟着一个以上的支撑句,以进一步具体说明将呈现的主题。研究方法、实验、程序、调查、计算、分析、结果以及其他重要的信息都将在这一部分出现。因此,这些支撑句可以被看作是摘要的"主体"部分。支撑句常用句式如下。

• The method used in our study is known as …

• The technique we applied is referred to as …

• The procedure they followed can be briefly described as …

• The approach adopted extensively is called …

• Detailed information has been acquired by the authors using …

• The research has recorded valuable data using the newly-developed method.

• This is a working theory which is based on the idea that …

• The fundamental feature of this theory is as follows.

• The theory is characterized by …

• The experiment consisted of three steps, which are described in …

• The test equipment which was used consisted of …

• Included in the experiment were …

• The winch is composed of the following main parts …

• We have carried out several sets of experiments to test the validity of …

• They undertook many experiments to support the hypothesis which …

• Recent experiments in this area suggested that …

• A number of experiments were performed to check …

• Examples with actual experiment demonstrate …

• Special mention is given here to …

• This formula is verified by …

• We also supply …

* **Concluding Sentence(s)**(结论句)

作为摘要的结尾部分,结论句通常用于分析结果、解释应用,并指出研究的重要性。同时,在写摘要时,我们应该选择适合具体文章的结构,并避免固守刻板的模式。结论句常用句式如下。

• In conclusion, we state that …

• In summing up it may be stated that …

• It is concluded that …

• The results of the experiment indicate that …

- The studies we have performed showed that …
- The pioneer studies that the authors attempted have indicated in …
- We carried out several studies which have demonstrated that …
- The research we have done suggests that …
- The investigation carried out by … has revealed that …
- Laboratory studies of … did not furnish any information about …
- All our preliminary results throw light on the nature of …
- As a result of our experiments, we concluded that …
- From our experiment, the authors came to realize that …
- This fruitful work gives explanation of …
- The author's pioneer work has contributed to our present understanding of …
- The research work has brought about a discovery of …
- These findings of the research have led the author to the conclusion that …
- The data obtained appear to be very similar to those reported earlier by …
- Our work involving studies of … prove to be encouraging.
- The author has satisfactorily come to the conclusion that …
- Finally, a summary is given of …

4. "五步法"撰写摘要("5 Steps" for Abstract Writing)

当动笔撰写摘要时,首先必须要熟悉论文的结构和内容。撰写摘要的主要步骤可概括为以下的"五步法":标记关键词和句子、列出论文要点、概括每个部分为一两句、起草摘要、检查终稿。

A. Underlining Key Words and Sentences

在论文关键词或关键句子下划线,尤其是一些表示过渡的词和句子。

B. Listing Essential Points of the Paper

仔细阅读并提取整篇论文要点,逐一列出。

C. Boiling down Each Section to a Sentence or Two

将每个部分概括为一两句话,这是准备摘要过程中一个有效地练习。

D. Drafting the Abstract

起草摘要,要尽可能重新组织简洁明了的句子。如果从文本中直接借用一些单词、短语或者整个句子,摘要可能冗长且不连贯。起草摘要时,要避免包含观点、例子、细节和解释。

E. Checking the Final Draft

在提交前进行最后的检查,确保摘要符合所有格式要求,并再次确认没有语法错误。

5. "5A 策略"("5A Strategy")

结合上述撰写摘要的"五步法",以下的"5A 策略"将进一步帮助大家撰写相对标准且理想的摘要。

通过回答以下 5 个问题,将帮助大家更容易完成一篇相对满意的摘要。

Q1:在学术领域,你研究的主题具有哪些共识?
Q2:论文将聚焦于哪个研究主题?
Q3:使用什么方法或材料来支持主要观点?
Q4:将得出什么结论?
Q5:论文的主要贡献是什么?

然后,可以通过使用以下公式来改进摘要。

"5A Strategy":Abstract=A1+A2+A3+A4+A5

Where, A1 to Q1 (one sentence)

A2 to Q2 (one topic sentence, one or two supporting sentences that set forth the research topic if necessary)

A3 to Q4 (two or three sentences to give specific information about the research)

A4 to Q4 (one sentence)

A5 to Q5 (one sentence)

For Example:①Modern management is essentially about managing people as well as processes in a rapidly changing environment. ②The author present the factors which make "strategic management" effective. ③A dominant factor is the organization climate which, in turn, is determined by the quality of the managers and the availability of alternatives. ④To improve the organization climate in which strategic management can be effective, the quality of the managers is a crucial factor. ⑤The scope for alternatives also proves an important constraint. ⑥The author suggests that the assessment of effects for management should include the use of consultants and the role of formal procedure. ⑦It is concluded that the correct judgment and optimal operation of the essential factors will enhance will effectiveness of strategic management in general.

A1:back ground—sentence ①;
A2:main topic—sentence ②;
A3:specific investigations—sentences ③④⑤;
A4:result & suggestion—sentence ⑥;
A5:conclusion & contribution—sentence ⑦.

6. 撰写摘要时常见的错误

A. Mixed Writing Style(混合写作文体)

专业或学术论文写作的特点是使用书面语,这与口语化风格的写作不同。具体示例如下。

Abstract:In this paper, <u>we have given a reason</u> that is why the competitive ability of the national firms is weak. <u>Because</u> the non-national firms can get very cheap labor, under the same technical and economical conditions and the same cost, a non-national firm can produce more output than a national firm does, <u>so</u> it can get much more profit. <u>In this way</u>, the competitive ability of the non-national firm is stronger than the national firms.

以上摘要示例较为口语化,类似于口头表达的风格。类似的表述,如 if you will, I think, my study has confirmed the saying that …

B. Over-simplified Statements(过于简单的表述)

尽管摘要短小且简洁,但不应过于简单,举例如下。

Abstract:In this paper, methods of tomato grafting are presented with an analyzed data by computer.

上面的摘要过于简略,未告知读者番茄嫁接是如何进行的、如何分析的以及结果究竟是什么等。示例的摘要只是标题的扩展,未能发挥微缩全文的作用。

Abstract:The authors in this paper present a process of tomato grafting innovation. The process is proved to be effective, and the authors are satisfied with results.

同样,以上摘要过于笼统和泛泛而谈,对核心内容即番茄嫁接技术创新过程未进行明确的介绍,同时对文章的结果和影响也缺乏明确的阐述,这篇摘要没有发挥出它基本功能。

C. Monotonous Expression(单调的表达)

在撰写摘要时,应追求形式上的多样性,如长句与短句交替使用,被动语态和主动语态、动词和短语的形态变化等,缺乏多样性可能导致表达单一化。

Exercise

Ⅰ. Mimicking the "5A Strategy" from the text, analyze the following abstract.

Abstract:(1)Tensile cracking-sliding loess collapse is the most recurrent disaster in the expressway construction in loess area. (2)The maximum vertical tension cracking depth in trailing edge of loess slope is obtained from the limiting equilibrium equation of soil; the constitutive relation between vertical fissure mass and the potential sliding surface mass is established by the water weakening function; the equation of balanced camber and cusp is formulated from the expression of total potential energy of the collapse, then the catastrophic model of cusp for tensile cracking-sliding loess collapse is established. (3)The formation mechanism for tensile cracking-sliding loess collapse is explored based on the mode. After excavation, the tension crack in trailing edge of loess slope is formed, and then the fissure is filled with water. (5) The water filling increases the shear modulus of vertical tension and decreases the shear modulus on the potential sliding surfaces. (6)When the saturation degree of loess mass reaches a certain value, the slope system undergoes mutation and the sliding surface is formed; then the core of the soil body moves unceasingly to the free surface. (7)Once the core of soil body sliding out of steep slope, collapse will be happen. (8)Based on the limiting equilibrium equation and the Mohr Coulomb's law, computing formula of the stability for cracking-sliding loess collapse is established. (9) The research results will be of valuable reference for the design and construction of loess collapse control.

II. Please identify the issues in the following two abstracts and then refine them.

Abstract: The influence of the high altitudes to growth of tomato and ... are discussed. Some design considerations of ... are given. A practical construction ... is shown. The method with high precision of ... is presented. The measured stability data ... are also given.

Abstract: In this paper, We obtain ... and using this, we give a necessary and sufficient condition ... For proving the above result, we introduce a new ... Furthermore, we prove ... Using this result, we get a necessary and sufficient condition that ... And finally, we discuss some properties of ...

Appendix Ⅰ Words and Expressions

A

abutting /əˈbʌtɪŋ/ adj. 邻接的,毗连的,紧靠的
actual fault 实际断层
adorment /əˈdɔːnmənt/ n. 装饰,装饰品
aeration /eəˈreɪʃ(ə)n/ n. 通风,通气
age-old 古老的,久远的
Alberta /ælˈbɜːtə/ n. 亚伯达(地名)
alive /əˈlaɪv/ adj. 活着的
Appalachian /ˌæpəˈleɪtʃiən/ n. 阿帕拉钦亚(地名)
asteroid /ˈæstərɔɪd/ n. 小行星
astrogeology /ˌæstrəʊdʒɪˈɒlədʒɪ/ n. 天体地质学
acute dihedral angle 锐二面角
alternating layer 互层

B

basalt /ˈbæsɔːlt/ n. 玄武岩
beach /biːtʃ/ n. 海(河、湖)滩
bedrock /ˈbedrɒk/ n. 基岩
biogeography /ˌbaɪəʊdʒɪˈɒgrəfi/ n. 生物地理学
boulder /ˈbəʊldə(r)/ n. 巨砾
breccia /ˈbretʃə; -tʃɪə/ n. 角砾岩
bedding plane 层面

C

calcite /ˈkælsaɪt/ n. 方解石
calcium carbonate n. 碳酸钙
Caledonian /ˌkælɪˈdəʊniən/ n. adj. 加里东期(的)
call upon 号召,指派
Cambrian /ˈkæmbriən/ n. adj. 寒武纪(的),寒武系(的)
cast off 抛弃
cavity /ˈkævəti/ n. 洞穴
cementing agent 胶结构
check /tʃek/ n. 制止,控制
clastic /ˌklæstɪk/ adj. 碎屑状的
clay /kleɪ/ n. 黏土
coarse /kɔːs/ adj. 粗粒的
cobble /ˈkɒbl/ n. 大砾,卵石,圆石
columnar jointing 柱状节理
comfort /ˈkʌmfət/ n. 舒适,使舒适的事物
compressibility /kəmˌpresɪˈbɪlɪtɪ/ n. 压缩性,压缩系数,压缩率
conglomerate /kənˈglɒmərət/ n. 砾岩
conjugate /ˈkɒndʒəgeɪt/ adj. 共轭的
convert into 转变成
country rock 围岩
crystallize /ˈkrɪstəlaɪz/ v. 使结晶
curvature /ˈkɜːvətʃə(r)/ n. 曲率
cylindrical /səˈlɪndrɪk(ə)l/ adj. 圆柱形的
chlorite /ˈklɔːraɪt/ n. 绿泥石
conjugate sets of shear joints 共轭剪节理组
cylindrical fold 圆柱形褶皱

D

debris /'debriː/   n. 碎片,岩屑
decay /dɪ'keɪ/   v. (使)腐烂
delineation /dɪˌlɪni'eɪʃn/   n. 描写,描绘
diabase /'daɪəbeɪs/   n. 辉绿岩
dihedral /daɪ'hiːdr(ə)l/   adj. 二面的
dike /daɪk/   n. 岩墙
dinosaur /'daɪnəsɔː(r)/   n. 恐龙
dip /dɪp/   n. 倾向
discharge /dɪs'tʃɑːdʒ/   v. 排泄,释放
disturb /dɪ'stɜːb/   v. 扰乱,弄乱
drought /draʊt/   n. 干旱季节,旱灾
durable /'djʊərəb(ə)l/   adj. 持久的,耐久的

E

ecology /i'kɒlədʒi/   n. 生态学
economic geology   经济地质学
elliptic /ɪ'lɪptɪk/   adj. 椭圆的
engineering geology   工程地质学
engineering property   工程特性
epitome   n. 节录,缩影,摘要
erosion /ɪ'rəʊʒ(ə)n/   n. 侵蚀
exert /ɪg'zɜːt/   v. 用(力)
extrude /ɪk'struːd/   v. 挤压出,喷出,突出
extrusive /ɪk'struːsɪv/   adj. 喷出的
eroded surface   侵蚀面
extrusive mass   喷出岩

F

facies /'feɪsiːz/   n. 相
feldspar /'feldspɑː(r)/   n. 长石
flakelike /fleɪklaɪk/   adj. 薄片似的,石片似的
fold /fəʊld/   n. 褶皱
fold axis   褶皱轴
fold-and-thrust belt   褶皱-逆冲带
foliar /'fəʊliə(r)/   adj. 叶的,叶状的
foliated /'fəʊlieɪtɪd/   adj. 叶片状
foliation /ˌfəʊli'eɪʃən/   n. 叶理,面理
footprint /'fʊtprɪnt/   n. 脚印
foundation /faʊn'deɪʃn/   n. 基础,地基
foundation engineering   基础工程,地基工程
fracture /'fræktʃə(r)/   n. 断裂
fragment /'frægmənt/   n. 碎片,碎屑
fragmental /fræg'mentəl/   adj. 碎屑的,破碎的
frame /freɪm/   v. 构成,限定

G

Garlock   n. 加洛克(地名)
gasoline /'gæsəliːn/   n. 汽油
gently /'dʒentli/   adv. 轻轻地
geochemistry /ˌdʒiːəʊ'kemɪstri/   n. 地球化学
geomorphology /dʒiːəʊmɔː'fɒlədʒi/   n. 地貌学
geophysics /ˌdʒiːəʊ'fɪzɪks/   n. 地球物理学
giant /'dʒaɪənt/   adj. 巨大的,庞大的
glaciology /ˌglæsi'ɒlədʒi/   n. 冰川学
gneiss /naɪs/   n. 片麻岩
granite /'grænɪt/   n. 花岗岩
gravel /'grævl/   n. 砾石,砂砾
ground water   地下水
gypsum /'dʒɪpsəm/   n. 石膏

H

haphazard /hæp'hæzəd/   adj. 偶然性,杂乱的
harden /'hɑːdn/   v. 使变硬,变坚固
heavy mineral   重矿物

hinge /hɪndʒ/   n. 枢纽
hornblende /ˈhɔːnblend/   n. 角闪石
hydrography /haɪˈdrɒɡrəfi/ n. 水文地理学
hydrology /haɪˈdrɒlədʒi/   n. 水文学

I
igneous /ˈɪɡniəs/   adj. 火的, 火成的
igneous rock   火成岩
illite /ˈɪlaɪt/   n. 伊利石
impart /ɪmˈpɑːt/   v. 给予, 传递
imprison /ɪmˈprɪzn/   v. 束缚, 堵塞
in time   及时, 按时, 准时
in view of   由于……, 鉴于……, 考虑到
insect /ˈɪnsekt/   n. 昆虫
integral /ˈɪntɪɡrəl/
     adj. 构成整体所必需的, 完整的
interfingering /ˈɪntəfɪŋɡərɪŋ/
     adj. 相应穿插的, 指状交叉的
intrude /ɪnˈtruːd/   v. 侵入
intrusive /ɪnˈtruːsɪv/   adj. 侵入的
intrusive mass   侵入岩
iron-bearing   含铁的

J
joint set   节理组

K
kaolinite /ˈkeɪəlɪˌnaɪt/   n. 高岭石
keep…from   使…免于

L
last /lɑːst/   v. 持续, 耐久
lava flow   熔岩流
left-lateral   左行的
limb /lɪm/   n. 翼
lime /laɪm/   n. 石灰
limestone /ˈlaɪmstəʊn/ n. 石灰岩, 碳酸钙
limnology /lɪmˈnɒlədʒi/   n. 湖沼学
limonite /ˈlaɪməˌnaɪt/   n. 褐铁矿
linear fabric   线状组构
lithologic /ˌlɪθəˈlɒdʒɪk/   adj. 岩性(学)的
loose /luːs/   adj. 松的, 松散的

M
magma /ˈmæɡmə/   n. 岩浆
magma chamber   岩浆层
mapmaking /ˈmæpmeɪkɪŋ/   n. 制图学
marble /ˈmɑːb(ə)l/   n. 大理岩
marine (submarine) geology
     海洋(海底)地质学
massive /ˈmæsɪv/   adj. 大块的
meteorite /ˈmiːtiəraɪt/   n. 陨星, 陨石
meteorology /ˌmiːtiəˈrɒlədʒi/   n. 气象学
methane /ˈmiːθeɪn/   n. 甲烷, 沼气
mineral /ˈmɪnərəl/   n. 矿物
mineralogist /ˌmɪnəˈrælədʒɪst/
     n. 矿物学家
mineralogy /ˌmɪnəˈrælədʒi/   n. 矿物学
mining geology   矿山地质学
miscellaneous /ˌmɪsəˈleɪniəs/
     adj. 混杂的, 各种各样的, 其他的
moisture /ˈmɔɪstʃə(r)/
     n. 潮湿, 湿气, 湿度
mold /məʊld/   n. 模子
mottling /ˈmɒtlɪŋ/ n. 斑点构造, 成斑作用
mudstone /ˈmʌdstəʊn/   n. 泥岩
metamorphism /ˌmetəˈmɔːfɪzəm/
     n. 变质作用
mica /ˈmaɪkə/   n. 云母

N
net slip   总滑距, 净滑距

O
occurrence /əˈkʌrəns/   n. 出现, 产状

oceanography /ˌəʊʃəˈnɒɡrəfi/ n. 海洋学
oozing /ˈuːzɪŋ/ v. 渗出，使（液体）缓缓流出
opening /ˈəʊpnɪŋ/ n. 孔，空隙
Ordovician /ˌɔːdəˈvɪʃən/ n. adj. 奥陶纪（的），奥陶系（的）
outcrop /ˈaʊtkrɒp/ n. 露头

P

paleontology /ˌpælɪɒnˈtɒlədʒi/ n. 古生物学
pebble /ˈpebl/ n. 细砾，卵石
pedology /pɪˈdɒlədʒi/ n. 土壤学
penetrate /ˈpenətreɪt/ v. 穿过，渗入，钻入
petrify /ˈpetrɪfaɪ/ v. （使）石化
petrology /pəˈtrɒlədʒi/ n. 岩石学
phyllite /ˈfɪlaɪt/ n. 千枚岩
physiography /ˌfɪziˈɒɡrəfi/ n. 地文学，自然地理学
pine /paɪn/ n. 松树
planar fabric 面状组构
planetary astronomy 行星天文学
plaster /ˈplɑːstə(r)/ n. 灰泥，熟石膏
plastic /ˈplæstɪk/ adj. 塑性的，塑造的
pond /pɒnd/ n. 池塘
pore /pɔː(r)/ n. 孔，孔隙
pore space 孔隙
porous /ˈpɔːrəs/ adj. 松散的
precipitation /prɪˌsɪpɪˈteɪʃn/ n. 沉淀，降雨量
pressing /ˈpresɪŋ/ adj. 紧迫的，追切的
principal curvature 主曲率
profile /ˈprəʊfaɪl/ n. 剖面（图），廓线，轮廓
profile section 剖面（图）
pyrite /ˈpaɪraɪt/ n. 黄铁矿

Q

quartz /kwɔːts/ n. 石英
quartzite /ˈkwɔːtsaɪt/ n. 石英岩

R

rainfall /ˈreɪnfɔːl/ n. 降水量，一场降雨
remake /ˈriːmeɪk/ v. 重制
render /ˈrendər/ v. 反映，表达
reptile /ˈreptaɪl/ n. 爬虫，爬虫类
resin /ˈrezɪn/ n. 松脂，松香
rock-forming 造岩的
rose diagram 玫瑰花图

S

sandstone /ˈsændstəʊn/ n. 砂岩
sandy /ˈsændi/ adj. 砂质的
Sarv n. 萨夫（地名）
saturate /ˈsætʃəreɪt/ v. 使饱和
saturation /ˌsætʃəˈreɪʃn/ n. 饱和，饱和作用
Scandinavian 斯堪的纳维亚（地名）
schist /ʃɪst/ n. 片岩
schistosity /ʃɪsˈtɒsɪti/ n. 片理
sediment /ˈsedɪmənt/ n. 沉积物，沉淀物
sedimentary rock 沉积岩
sedimentation /ˌsedɪmenˈteɪʃn/ n. 沉积物形成作用，沉积作用
seething /ˈsiːðɪŋ/ adj. 炽热的，沸腾的
shale /ʃeɪl/ n. 页岩
shear joint 节理
shell /ʃel/ n. 贝壳
shore /ʃɔː(r)/ n. 岸
shoreline /ˈʃɔːlaɪn/ n. 海岸线
silicate /ˈsɪlɪkeɪt/ n. 硅酸盐
silicate melt 硅酸盐熔融体
sill /sɪl/ n. 岩床，海底山岩

silt /sɪlt/	n. 粉砂
siltstone /ˈsɪltstəʊn/	n. 粉砂岩
simultaneously /ˌsɪmlˈteɪniəsli/	adv. 同时地
single-layer /ˈsɪŋ(ə)lˈleɪə(r)/	单层的
slab /slæb/	n. 板片
slate /sleɪt/	adj. 板岩
slaty cleavage	板劈理
smectite /ˈsmektaɪt/	n. 蒙脱石
snail /sneɪl/	n. 蜗牛
soak /səʊk/	v. 浸, 渗透
soak into	渗入
solidify /səˈlɪdɪfaɪ/	v. 使固化, 使凝固
spacing /ˈspeɪsɪŋ/	n. 间距, 间隔
spherical projection	球面投影
sticky /ˈstɪki/	adj. 黏的
still-liquid	仍处于液体的
stratigraphic discontinuity	地层间断
stratigraphic omission	地层缺失
stratigraphic repetition	地层重复
stratigraphy /strəˈtɪɡrəfi/	n. 地层学
stream /striːm/	n. 小河, 川, 溪流
strength /streŋθ/	n. 强度
stretching /ˈstretʃ/	n. 伸展, 伸长
strike /straɪk/	n. 走向
string out	使成串地展开, 连成一列
structural geology	构造地质学
sulfur /ˈsʌlfə(r)/	n. 硫
surrounding rock	围岩
susceptible /səˈseptəbl/	adj. 受……影响的
swamp /swɒmp/	n. 沼泽, 沼泽地
swirling /ˈswɜːl/	adj. 旋转的
syndepositional /sɪndɪpɒˈzɪʃnl/	adj. 同沉积(作用)的
syndepositional faulting	同沉积断层
systematic joint	系统节理

T

tabular /ˈtæbjələ(r)/	adj. 板状的
talc /tælk/	n. 滑石
tanks /tæks/	n. 贮水池; 大容器
tectonically /tekˈtɒnɪk/	adv. 构造上地
Tennessee	n. 田纳西(地名)
three-phase structure	三相结构
thrust /θrʌst/	n. 逆断层, 逆冲断层
thrust out	挤出, 接出
thrust sheet	逆掩盘
topography /təˈpɒɡrəfi/	n. 地形学
trace /treɪs/	n. 痕迹
transpiration /ˌtrænspɪˈreɪʃn/	n. 蒸发, 蒸腾作用
trap /træp/	v. 使陷入, 圈闭
truncate /trʌŋˈkeɪt/	v. 截去……的顶端, 截断, 截短

U

Upper Precambrian	上前寒武纪(的)
uppermost /ˈʌpəməʊst/	adj. 最上的

V

vadose /ˈveɪdəʊs/	adj. 渗流的
vadose water	循环水, 渗流水
vague /veɪɡ/	adj. 模糊的
vein /veɪn/	n. 脉, 岩脉
vertical joint	垂直节理
volcanic vent	火山口

W

wall-rock	围岩
wall-rock alteration	围岩蚀变
wander /ˈwɒndə(r)/	v. 徘徊
wash /wɒʃ/	v. 洗, 冲刷

water table	潜水面	weathering /ˈweðərɪŋ/	n. 风化作用
wavy /ˈweɪvi/	adj. 波状的,起伏的	wet /wet/	adj. 潮湿的
wavy foliation	波状叶理		
weather /ˈweðə(r)/	v. (使)风化	zone of aeration	包气带

Z

附:地史单位表

The Geologic History Unit Table

时间(年代)地层单位 Time stratigraphic unit			地质(年代)时代单位 Geologic era unit	
宇 Eonthem			宙 Eon	
界 Erathem			代 Era	
系 System			纪 Period	
统 Series	上 Upper		世 Epoch	晚 Late
	中 Middle			中 Middle
	下 Lower			早 Early
阶 Stage			期 Age	
时带 Chronozone			时 Chron	

Appendix Ⅱ Plate

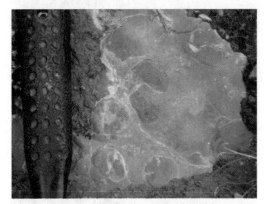

biogenic limestone

（生物灰岩）

fuchsia pelitic siltstone

（紫红色泥质粉砂岩）

parallel bedding

（平行层理）

bidirectional cross bedding

（双向交错层理）

sand ripple cross bedding

（沙纹交错层理）

tabular cross bedding

（板状交错层理）

bioglyph
（生物遗迹）

biological shell
（生物壳）

stylolite
（缝合线）

augen structure
（眼球状构造）

tilted stratum
（倾斜岩层）

parall unconformity/disconformity
（平行不整合接触）

Appendix II Plate

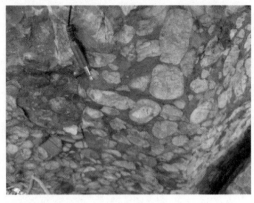

conglomerate

(砾岩)

lacustrine deposit

(湖相沉积)

fold

(褶皱)

normal fault

(正断层)

vertical joint

(垂直节理)

lenticle

(透镜体)

flowing area of mud flow
（泥石流流通区）

waterfall
（瀑布）

wind erosion hole
（风蚀洞）

geological process of contemporary galcier
（现代冰川地质作用）

sedimentation and desert
（沉积作用与沙漠）

piedmont alluvial plain
（山前冲积平原）

modern salt lake sedimentary deposit

（现代盐湖沉积矿床）

sand dune

（沙丘）

eolian erosion and sand ripple

（风蚀作用与波纹）

eolian erosion landform

（风蚀地貌）

lake erosion

（湖水侵蚀作用）

talus

（倒石锥）

Danxia landform
（丹霞地貌）

physical and chemical weathering of piedmont rock mass（山前岩体物理风化作用）

downcutting
（下切作用）

vertical erosion
（下蚀作用）

cataclastic rock mass
（碎裂岩体）

quartz vein
（石英脉）

Appendix Ⅱ Plate

different weathering

(差异风化)

limestone

(石灰岩)

fluvial terrace

(河流阶地)

bank erosion

(岸边侵蚀)

palaeobios of Qaidam Basin

(柴达木盆地古生物)

Kanbula National Geopark

(坎布拉国家地质公园)

paleontological fossil profile

（古生物化石剖面）

lacustrine deposit of Qinghai Lake

（青海湖湖相沉积）

production practice area of Zhamashan

（扎麻山生产实习区）

production practice area of Wulonggou

（五龙沟生产实习区）

field sampling and measuring

（现场取样与测量）

footrill sampling and measuring

（平硐取样与测量）